THE ESSENCE OF

# POWER ELECTRONICS

# THE ESSENCE OF ENGINEERING SERIES

**Published titles**
The Essence of Solid-State Electronics
The Essence of Electric Power Systems
The Essence of Measurement
The Essence of Engineering Thermodynamics
The Essence of Analog Electronics

**Forthcoming titles**
The Essence of Optoelectronics
The Essence of Microprocessor Engineering
The Essence of Communications
The Essence of Digital Design

THE ESSENCE OF

# POWER ELECTRONICS

**J.N. Ross**
University of Southampton

**Prentice Hall**
LONDON   NEW YORK   TORONTO   SYDNEY   TOKYO
SINGAPORE   MADRID   MEXICO CITY   MUNICH   PARIS

First published 1997 by
Prentice Hall Europe
Campus 400, Maylands Avenue
Hemel Hempstead
Hertfordshire, HP2 7EZ
A division of
Simon & Schuster International Group

Typeset in 10/12pt Times
by MHL Typesetting Ltd, Coventry

Printed and bound in Great Britain by
Hartnolls Limited, Bodmin, Cornwall

---

Library of Congress Cataloging-in-Publication Data

---

Ross, J.N.
    The essence of power electronics / by J.N. Ross.
        p.     cm. – (The essence of engineering)
    Includes bibliographical references and index.
    ISBN 0-13-525643-7 (pbk. : alk. paper)
  1. Power electronics.   I. Title.   II.   Series.
  TK7881.15.R67    1997
621.31′7–dc20                                            95-25452
                                                              CIP

---

British Library Cataloguing in Publication Data

---

A catalogue record for this book is available from
the British Library

ISBN 0-13-525643-7

---

1   2   3   4   5      01   00   99   98   97

# Contents

# Preface

The scope of power electronics is very wide, and the power levels controlled range from hundreds of megawatts for converters embedded into power transmission systems down to a few watts for the control of drive motors in domestic audio equipment. With such a wide spectrum of power, it is impossible to do justice to the whole subject in a short text such as this. Here the aim is to approach the subject from the low-power end of this spectrum. This choice is reflected in the applications considered, for example the discussion of controlled rectifiers is restricted to single-phase circuits. The book is directed at students studying electronic engineering, while many of the excellent texts available on power electronics approach the subject from the viewpoint of the electrical power engineer. These texts have not always proved very accessible for students, whose interests are perhaps more centred on computers and electronic equipment.

The book arose from a single-semester course on power electronics, which was taught as a third-year option, to students on a three-year Bachelor of Engineering or a four-year Master of Engineering degree. The course lasted for a single semester (24 lectures), and its aim was to give the students an overview of the subject and, if possible, interest them in it. This philosophy has been carried over into this work. Excessive use of mathematics has been avoided, although an attempt has been made to illustrate some of the different analytic techniques that may be used to analyse circuits that are generally non-linear and frequently use switching techniques.

The first three chapters deal with aspects of the basic technology. The first chapter provides an overview of the most important semiconductor devices which form the basis of all modern power electronic systems. The chapter assumes that the student has a basic understanding of how junction diodes, bipolar transistors and field effect transistors work. A detailed knowledge of the physics is not required. The aim of the chapter is to draw out the important characteristics of the power versions of these devices, especially with regard to their use as power switches. The second chapter looks at the dissipation of heat from power electronic devices, and discusses the simple models used for thermal calculations. The third chapter reviews magnetic components, i.e. inductors and transformers. These components are largely avoided in low-power electronic systems, and hence, the students are not very familiar with them. However, for power electronics an understanding of their properties and behaviour is essential.

The remaining nine chapters examine some of the techniques used in power electronics as applied to a range of applications. The emphasis is on power supplies, power converters and power regulators. The techniques discussed are primarily those used in electronic equipment, where the power rarely exceeds 1 kW, although essentially the same techniques are used at much higher power. Much emphasis is placed on switching techniques, which have become very widespread because of their high conversion efficiency.

Most power electronic systems involve the use of feedback in the control system. It is not possible within the constraints of a book like this to consider control of switching power controllers and converters in detail, and many texts avoid the issue altogether. However, the design of the power stage will usually be strongly influenced by the need to enclose it within a feedback loop, and thus it is not really possible to consider the power stage without taking into account the needs of the control system. In order to illustrate some of the control issues, two design studies have been put together to form Chapter 8. These look at a range of issues that need to be considered in the design of practical switching regulators, and one of the most significant of these issues is control. This chapter requires the reader have some basic knowledge of control theory, and to understand the significance of poles and zeros in determining the transfer function of a system. However, heavy use of mathematics is avoided by the use of a graphical (Bode diagram) approach to feedback loop design. If necessary this chapter could be skipped.

Chapter 11 examines the use of power electronics to control small electric motors of the type that may be encountered in servo-systems, computer disk drives, video recorders and similar equipment. The final chapter, on audio power amplifiers, covers a topic not usually included in a book on power electronics. It is, however, a topic that is frequently of interest to students, and it relates well to other topics of linear and switching regulators. A power amplifier is only a rather specialized power converter!

Circuit simulation has been used to investigate some of the circuits described and illustrate their behaviour. The aim here is to illustrate the behaviour of the circuits under analysis and how circuit analysis may be used to analyze power circuits, and to encourage the student to experiment. The simulator used was the Evaluation Version of PSPICE$^{TM}$, from the MicroSim Corporation, running on an IBM-compatible PC. The source netlists are given, together with any device models used in the simulations. It should be possible to run the simulations on other packages with only minor changes. Simulation is a useful aid to understanding, especially of non-linear circuits, but it is only really effective if used alongside analytic techniques, to complement them, not replace them.

On a more personal note, I am grateful to my colleagues at the University of Southampton for many useful discussions, over many cups of coffee, and to the students who have listened to my lectures, and unwittingly contributed to this book with their stimulating questions. Finally, I would like to acknowledge the support of my wife and family for their patience while this book was in preparation.

# Symbols

This list contains the main symbols used within the text. In places other subscripts are used to identify particular variables. As with all the symbols, these are explained where they are used.

| | |
|---|---|
| $A$ | Area or constant of integration |
| $A_L$ | Inductance factor (specific inductance) |
| $A_V$ | Voltage gain |
| $B$ | Magnetic flux density |
| $B_r$ | Remanence |
| $C$ | Capacitance |
| $C_n$ | Amplitude of $n$th harmonic, cosine component in Fourier series |
| $C_\theta$ | Thermal capacity |
| $D$ | Duty ratio |
| $E$ | Energy or back-e.m.f. of motor |
| $f$ | Frequency |
| $G_m$ | Forward transconductance (of a transconductance gain stage) |
| $g$ | Length of air gap in magnetic core |
| $g_m$ | Forward transconductance (of a transistor) |
| $H$ | Magnetic field strength |
| $H_C$ | Coercivity |
| $h_{FE}$ | Common emitter current gain for bipolar transistor |
| $H(s)$ | Transfer function |
| $I$ | Electric current, constant or average value |
| $i$ | Electric current, varying with time |
| $K_E$ | E.M.F. constant for electric motor |
| $K_T$ | Torque constant for electric motor |
| $k$ | Boltzmann's constant ($1.3807 \times 10^{-24}$ J K$^{-1}$) |
| $L$ | Inductance |
| $N$ | Number of turns |
| $n$ | Turns ratio for transformer, or ideality factor for junction diode, or integer |
| $P$ | Power |
| $Q$ | Electric charge or '$Q$' factor for resonant circuit |
| $q$ | Electric charge or charge of an electron ($1.602 \times 10^{-19}$ J) |
| $R$ | Resistance |

| | |
|---|---|
| $R_\theta$ | Thermal resistance |
| $\mathfrak{R}$ | Reluctance |
| $S$ | Stabilization factor |
| $S_n$ | Amplitude of $n$th harmonic, sine component in Fourier series |
| $T$ | Period, or temperature, or torque |
| $t$ | Time |
| $t_f$ | Current fall time (in switch) |
| $t_r$ | Current rise time (in switch) |
| $V$ | Voltage, constant or average |
| $V_T$ | Thermal voltage $(kT/q)$ |
| $v$ | Voltage, varying with time |
| $Z$ | Impedance |
| $Z_{\theta jc}$ | Thermal impedance from junction to case for a transistor |
| $\alpha$ | Thyristor firing angle, or phase angle |
| $\beta$ | Common emitter current gain for bipolar transistor, or current extinction angle for thyristor converter |
| $\gamma$ | Overlap angle |
| $\lambda$ | Thermal conductivity |
| $\mu$ | Permeability |
| $\mu_o$ | Permeability of free space $(1.257 \times 10^{-6} \text{ H m}^{-1})$ |
| $\mu_r$ | Relative permeability |
| $\rho$ | Resistivity or slope resistance of shunt regulator |
| $\tau$ | Time constant |
| $\Phi$ | Magnetic flux |
| $\phi$ | Angle or phase displacement angle |
| $\Omega_R$ | Rotational angular velocity |
| $\Omega_S$ | Synchronous angular velocity |
| $\Omega_{SL}$ | Slip speed for synchronous motor |
| $\omega$ | Angular frequency |

CHAPTER 1

# Semiconductor devices

## 1.1 Introduction

Power electronics is concerned with the control and conversion of electrical power using electronic techniques. It covers a very wide range of power levels from a few watts, as, for example, in the power supplies for electronic equipment, to gigawatts for large d.c. transmission schemes. Many functions in the control and conversion of electrical power that were once performed using rotating machines now use 'static converters' based upon semiconductor devices used as switches.

That power electronics has developed rapidly in recent years is largely due to the development of solid-state electronic devices that are able to switch large currents and withstand large voltages. The technology is still developing rapidly and in this chapter only the basic and well-established devices will be considered. The objective is to provide the reader with sufficient knowledge of the devices and their characteristics to understand the circuits and techniques to be described in later chapters. It is assumed that the reader is familiar with the operation of basic bipolar and field effect semiconductor devices such as the junction diode, bipolar transistor and MOS transistor. However, a detailed knowledge of the semiconductor physics is not necessary.

For the reader wishing to learn more about power devices, Mohan, Undeland and Robbins (1995) give a good overview, with details of structures and an outline of the basic physics for all the main current device types. For a more detailed description of the device physics and fabrication techniques see Jayant Baliga (1987).

## 1.2 Power semiconductor diodes

### 1.2.1 Junction semiconductor diodes

The most widely used semiconductor rectifier diodes are based on the silicon $p$–$n$ junction. The voltage current characteristics of these diodes generally obey the ideal diode equation:

$$I_D = I_S \left( \exp\left(\frac{V_D}{nV_T}\right) - 1 \right) \tag{1.1}$$

where $I_D$ is the diode current, $V_D$ the diode voltage, $I_S$ the saturation current, $V_T$ is the thermal voltage ($kT/q$) and $n$ is an empirical constant known as the *emission coefficient* or *ideality factor*, whose value lies between 1 and 2. For power diodes the contact and substrate resistances cannot generally be ignored and a series resistance term must be considered as well as the ideal diode. For real devices, the reverse leakage current is generally greater than predicted by the Schockley equation. This is a consequence of the simplified theoretical models used to derive the equation.

The most important ratings that apply to a power diode are for reverse voltage and forward current. The reverse voltage rating normally specified is the *repetitive reverse maximum* voltage, $V_{RRM}$. This is less than the reverse breakdown voltage and short transient voltages of slightly higher value may occasionally be tolerated under fault conditions. For the forward current there are generally several ratings, and the *maximum average forward* current, $I_{F(AV)}$ and the *maximum r.m.s. forward* current, $I_{F(RMS)}$ are usually the most important. There may also be a *maximum surge* current specified typically for a single half cycle at power frequency. For reliable operation none of the ratings should be exceeded.

### 1.2.2 Reverse recovery

For a junction diode carrying current in the forward direction minority carriers are injected into the regions on either side of the junction. These carriers represent a stored charge that can normally only be removed either by recombination or by drift to the electrodes. Recombination is usually the more significant mechanism. For the diode to turn off when the direction of current is reversed these minority carriers must be removed to allow the depletion layer to develop. The diode is unable to block the reverse current immediately, but continues to conduct for a short period, while the charges recombine, giving rise to the reverse recovery transient. Figure 1.1 shows typical reverse recovery transients as the current is ramped down. The recovery is said to be *soft* if the reverse current decays relatively slowly to zero, while it is *abrupt* if the current falls quickly. Soft recovery diodes are generally used in switching circuits to avoid large, fast, voltage transients. For a particular diode, the total recovery time $t_{rr}$ is independent of the rate at which the current ramps down (at least to a first approximation) while the peak reverse current increases with an increasing rate of decrease of current.

Diodes are classified according to their reverse recovery times. A fast recovery diode may have a $t_{rr}$ of order 30–100 ns (somewhat larger at very high current ratings), while a diode for use at power frequencies may have a $t_{rr}$ of 10 $\mu$s or more. The use of a slow-recovery diode in a fast switching circuit results in excessive power dissipation in the diode and large transient currents that may cause excessive electromagnetic interference.

### 1.2.3 Reverse breakdown

The breakdown mechanism in power rectifier diodes is generally avalanche multiplication. As the reverse voltage approaches the breakdown voltage the

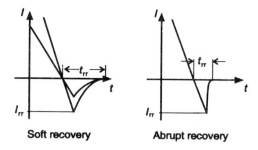

**Figure 1.1** *Reverse recovery transients for semiconductor junction diodes*

reverse leakage current increases rapidly (Figure 1.2). If the power dissipation is limited reverse breakdown is not necessarily destructive. A voltage reference or Zener diode is designed to operate at breakdown, with the current limited by a series resistance. Another use of reverse breakdown in a diode is to limit the voltage across an inductor, and to dissipate stored energy, when switching the current. Some diodes are designed and rated for this type of application.

### 1.2.4 Schottky diodes

A Schottky diode is formed between a semiconductor and a metal layer. The charge storage that limits the reverse recovery of junction diodes does not occur in Schottky diodes and reverse recovery is very fast. Schottky diodes are characterized by a lower forward voltage than silicon junction diodes and a larger reverse leakage current. They are readily available with current ratings of less than 100 A and reverse voltage ratings up to about 50 V. With their low forward voltage and rapid reverse recovery they are widely used in low-voltage switching power supplies. The current-voltage characteristic of a forward biased Schottky diode is of the same form as for a junction diode (equation (1.1)).

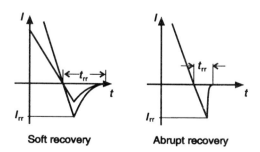

**Figure 1.2** *Voltage current characteristic for a semiconductor diode showing forward and reverse currents*

## 1.3  Bipolar junction transistors

### 1.3.1  Structure

The structure of a power bipolar junction transistor (BJT) does not differ significantly from that of a low-power device, except that the area is greater. A cross-section through a typical $n$–$p$–$n$ power transistor is shown in Figure 1.3. The primary requirements for a power transistor as opposed to a small signal transistor are a high voltage rating for the collector to base or emitter and a large collector current rating. The large voltage rating is achieved by using low doping in the collector region to increase the width of the depletion layer of the base collector junction, and hence reduce the electric field across this junction. The large collector current rating is achieved by having a large device area.

### 1.3.2  Characteristics

Power bipolar transistors generally have characteristics far from those of the ideal transistor. The common emitter current gain ($h_{FE}$) is generally low, falling at low and high currents. A typical power transistor might have a current gain of about 10 at its maximum current. High-frequency performance is also poor, with a typical unity gain frequency ($f_T$) of only a few MHz. Usually of more importance for power electronics is the current rise or fall time when the device is used as a switch, which is typically of order 0.3–1 $\mu$s for devices able to dissipate moderate power.

When used as a switch, as is usually the case in power electronics, the transistor may be driven into saturation (both the base emitter and base collector junctions forward biased). A saturated BJT gives a low voltage drop between collector and emitter (typically much less than 1 V). However, a saturated BJT is slower to turn off than one not in saturation. The collector current will not start to fall as soon as the base current is removed, but there will be a significant delay (the *storage delay*). This arises from excess charge stored in the base region which must be removed before the transistor can start to turn off. The storage delay may be minimized by imposing a reverse base current at turn-off, to help remove this stored charge.

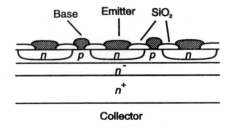

**Figure 1.3**  *The cross-sectional diagram of a bipolar power transistor*

### 1.3.3 Ratings

The maximum working voltage of a bipolar transistor is limited by the breakdown voltage of the base collector junction. There are generally two ratings, $V_{CEO}$, the maximum collector–emitter voltage with the base open circuit, and $V_{CBO}$, the maximum collector–base voltage with the emitter open circuit. $V_{CBO}$ exceeds $V_{CEO}$. With a finite resistance between the base and emitter, the maximum collector–emitter voltage will lie between these two values. The maximum permissible voltage can therefore sometimes be increased by ensuring the base is driven from a low source impedance, but generally it is wise to take $V_{CEO}$ as the maximum voltage.

The collector current is the subject of a maximum continuous rating, and sometimes a surge rating for short pulses with a low duty cycle. In addition, there will be limits imposed by the power dissipation. All these limits are conveniently brought together in the *safe operating area* diagram. This is based on a logarithmic plot of collector current against collector voltage, as shown in Figure 1.4. The solid line shows the bounding values for d.c. operation. The safe operating area (SOA), in which the device is within safe operating conditions, is the area to the lower left of the bounding line. The voltage-, current- and power-limiting lines are shown. In addition, for bipolar transistors there is a limiting line with a slope of $-2$ on the log-log plot which is due to the second breakdown phenomenon. Second breakdown occurs at high collector voltage, and results from localized regions of high current density within the transistor, which heat up, leading to local *thermal runaway*, where the temperature continues to rise out of control. This rapidly leads to the destruction of the transistor. The SOA, as with all ratings, is at a specified case temperature, often 25°C. At higher temperatures the power limit must be reduced.

When the transistor is operated so that significant power is dissipated only for short intervals, then the safe operating area may be extended as shown by the dotted line. The position of the boundary is determined by the duration of the pulse(s), $t_w$,

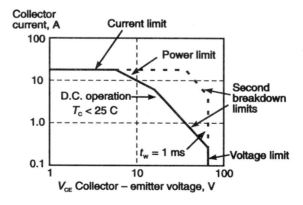

**Figure 1.4** *The safe operating area for a power bipolar transistor*

and, for a train of pulses, by their repetition rate. For very short times (typically of order $10\,\mu s$) the area is bounded by the current and voltage limits only.

### 1.3.4 Driving bipolar transistors

A BJT switching a large current needs a base drive current sufficient to keep the transistor turned on. The current gain may be quite low, so the base drive may well be 10% of the load current. At turn-off there will be a large stored charge in the base that must be removed. It is therefore desirable that a reverse voltage is applied to the base at turn-off to help remove this charge by conduction through the base connection. In addition, to ensure a rapid turn-on, it is desirable to provide an increased base drive for a short time so as to increase the rate at which the base is charged. Figure 1.5(a) shows a simple circuit which may be used to provide a suitable drive. The circuit is driven from a voltage source which switches between 0 and $V$ volts. The bias current is limited by $V$ and $R1 + R2$, with an initial current surge to charge up the base limited in magnitude by $R1$ and duration by $C$. At turn off the charge in the capacitor C drives the base negative and helps the recovery.

To ensure fast switching it is better to avoid the BJT saturating. This means that the base drive must ensure that the collector-to-emitter voltage is kept as low as possible (to minimize losses), while not allowing the collector base junction to become forward biased. One way of achieving this is to use a Baker's clamp (Figure 1.5(b)). The diode D1 will conduct if the collector voltage falls below the base voltage, thus limiting the base current to that needed to reduce the collector–base voltage to about zero. The diode D2 provides an offset voltage to allow for the voltage drop across D1, while diode D3 provides a path for applying a reverse base drive.

### EXAMPLE 1.1

A bipolar transistor operating as a saturated switch needs a base drive of 0.2 A to ensure saturation under all load conditions. The circuit shown in Figure 1.5(a) is to be used to provide the base drive, with a source voltage

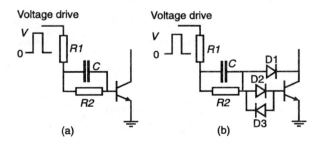

**Figure 1.5** *Base drive circuits: (a) simple; (b) with a Baker's clamp*

of 0 V in the *off* state and 10 V in the *on* state. The peak forward base drive current at turn on should be about 0.5 A and the time constant for the peaking network should be about $2 \mu s$. Choose suitable values for $R_1$, $R_2$ and $C$.

**SOLUTION**

Assume that the base–emitter voltage is about 0.7 V. Hence the steady-state base current is given by

$$I_B = \frac{10 - 0.7}{R_1 + R_2}$$

If $I_B$ is 0.2 A, then $R_1 + R_2 = 46.5 \, \Omega$. The peak base current is 0.5 A and the current is limited only by $R_1$, hence

$$R_1 = (10 - 0.7)/0.5 = 18.6 \Omega$$

Choosing the nearest standard values gives $R_1$ equal to $18 \, \Omega$ and $R_2$ equal to $27 \, \Omega$.

During turn-on, and during the storage time at turn-off, the base voltage will be almost constant. Thus to deduce the time constant $R_1$ and $R_2$ must be considered in parallel to give the equivalent Thevenin resistance for charging or discharging $C$. Thus the time constant is given by

$$\tau = \frac{C R_1 R_2}{R_1 + R_2}$$

If $\tau$ is $2 \mu s$ this gives $C$ equal to 185 nF, so taking the nearest value $C$ should be 180 nF.

Note that with this large base current the power dissipation in the base resistors must be considered. In the steady state, with the transistor *on* the power dissipation will be 0.77 W in $R_1$ and 1.15 W in $R_2$. If the switch is turned on and off rapidly then the mean power will be reduced by the fraction of the time for which the switch is *on*, but the power dissipated in charging and discharging $C$ must also be considered.

## 1.4 MOSFETs

### 1.4.1 Structure

Low-power MOSFETs use a *lateral* structure in which the current flow is parallel to the surface of the silicon die, with both source and drain on the upper surface. While this structure can be scaled up to handle larger power, it makes inefficient use of the silicon area. Power MOSFETs usually use a *vertical* DMOS (diffused MOS) structure. A DMOS transistor is fabricated as a large number of small cells, all connected in parallel. A cross-section through two of these cells, for an *n-*

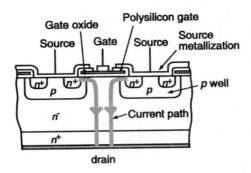

**Figure 1.6** *The cross-sectional diagram of a DMOS power transistor*

channel device, is shown in Figure 1.6. Each cell is based upon a p-type region diffused into the silicon die. The n-type source region is then diffused into this p-type well. The channel of the FET is formed at the surface of the silicon, beneath the gate and around the periphery of the cell. The device is an enhancement MOSFET, so with zero gate voltage no source drain current is able to flow. Applying a positive gate voltage creates the channel and enables the current to flow from the source region to the drain. The source connection is on top of the die while the drain is beneath it, so that the current flow is vertical.

P-channel power MOSFETs are available, with similar structure but the n and p layers interchanged. The majority carriers in a p-channel (holes) have lower mobility than the electrons in an n-channel. A consequence of this is that p channel devices need a greater channel width, and hence a larger device area, for the same transconductance and on state resistance. The larger area gives greater gate source and gate drain capacitance and increases the device cost. Generally, n-channel devices are preferred where possible.

With the DMOS structure there is a diode, the *body diode*, between the source and the drain. This diode conducts when the drain–source voltage is negative for an n-channel MOSFET or if this voltage is positive for a p-channel MOSFET. In some power MOSFETs this diode is configured so that it may be used as a reverse diode across the transistor, in which case its characteristics are specified in the data sheet. Otherwise it is advisable to ensure that the body diode never conducts.

### 1.4.2 Characteristics

DMOS power MOSFETs are enhancement devices, and the source–drain current is zero if the gate–source voltage, $V_{GS}$, is less than the threshold voltage, $V_{GS(th)}$. The simple model for an enhancement FET predicts that the drain current should increase quadratically with $V_{GS}$ for voltages above $V_{GS(th)}$, assuming that the drain–source voltage is large enough that the channel is pinched off and the drain current is saturated. For power MOSFETs this behaviour is observed just above threshold, but at larger drain currents the relation between drain current and gate voltage becomes linear.

When considering a power MOSFET as a switch, in the on-state when the drain voltage is small the FET channel behaves as a resistance. The channel resistance decreases as the gate voltage increases, but the minimum value of the on-state resistance $R_{DS(on)}$, is limited by the resistance of the $n$ layer through which the current must flow to the drain terminal. The higher the voltage rating of the transistor, the lower must be the doping of this layer, and the higher $R_{DS(on)}$.

The gate–source capacitance $C_{GS}$ and gate–drain capacitance $C_{GD}$ of a power MOSFET are large. For an $n$-channel MOSFET rated at 10 A, $C_{GS}$ is of order 1 nF and $C_{GD}$ of order 150 pF when $V_{GS}$ is zero. Values are greater for $p$-channel devices. Hence, when the transistor is used as a switch a significant charge must be supplied to the gate to turn the device on or off. It is therefore convenient to consider the device to be controlled by the gate charge. The manufacturer's data sheet often shows a graph of gate voltage against gate charge (Figure 1.7) when switching a specified load resistance across a specified supply voltage. This graph may be used to estimate turn-on or turn-off delay time and the current rise or fall time. The graph may be divided into three regions. In region I the transistor is off, and the gate voltage rises with charge. In region II the transistor is turning on, and the gate–drain capacitance is charged as the drain voltage falls. In region II the gate voltage changes only slowly. This is a manifestation of the Miller effect, giving rise to a very large effective input capacitance. In region III the transistor is on, the drain voltage is close to zero, and as more charge is supplied to the gate, the gate voltage continues to rise. If the source voltage and source impedance of the gate drive circuit are known, this graph of gate voltage against gate charge may be used to estimate the rate of rise or fall of current in the transistor.

### 1.4.3 Ratings

As with the BJT, the ratings of a MOSFET are conveniently collected together onto a safe operating area diagram. This is somewhat simpler than for a BJT since second breakdown does not occur in a majority carrier device. A typical example of

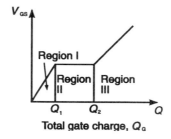

**Figure 1.7** *Gate charge characteristics of a power MOSFET*

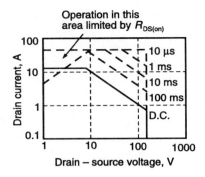

**Figure 1.8** *The safe operating area of a power MOSFET*

the SOA for a power MOSFET is shown in Figure 1.8. The area is bounded by the maximum current, the power dissipation (the line with a slope of −1), and the maximum voltage. The peak current exceeds the maximum continuous current by a large factor, even for quite long pulses. At high current the minimum drain voltage is limited by the on-state resistance, $R_{DS(on)}$, which may be very low for low-voltage devices, but increases rapidly for devices rated at a few hundred volts.

Another important rating for a MOSFET is the maximum gate source voltage. This is obviously limited by the breakdown voltage of the gate oxide. A typical value for the maximum gate source voltage is ±20 V. Breakdown of the gate oxide will result in permanent damage, so that care must be exercised in handling the devices to avoid damage from static electricity.

### 1.4.4  Driving MOSFET switches

To turn a MOSFET on or off quickly requires the gate charge to be changed as rapidly as possible. This is generally achieved by driving the gate from a low-impedance voltage source which can source and sink current. To turn on a 10 A MOSFET may require a gate charge of about 20 nC. Thus if a turn-on (or turn-off) time of around 100 ns is required a current of ±0.2 A is necessary with a voltage swing of about 0–10 V. The driver must also have a rise time which is comparable with the required turn-on (or turn-off) time for the MOSFET. This is a non-trivial requirement, although several manufacturers produce ICs specifically for driving MOSFET switches (e.g. Unitrode UC1705). A simple circuit capable of reasonable performance is shown in Figure 1.9, which uses two transistors in a totem-pole configuration.

### EXAMPLE 1.2

A MOSFET switches the current in a load. The data sheet for the MOSFET provides a plot of gate–source voltage against gate charge similar to that shown in Figure 1.7. The values of $Q_1$ and $Q_2$ are 7 nC and

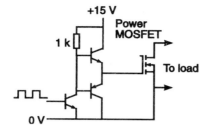

**Figure 1.9** *A gate drive circuit for power MOSFETs*

14 nC, respectively, and the flat region where the device turns on occurs at
a gate voltage of 7 V. The gate driver may be represented by a Thevenin
equivalent circuit, comprising a voltage source which rises as a sharp step
from 0 to 10 V and has a source resistance of 50 Ω. Estimate the time
between the voltage step and the time at which the transistor is fully turned
*on*.

## SOLUTION

The process of charging the gate can be considered in two phases. In
region I the input is a capacitance and the gate will charge exponentially.
In region II the charging occurs at constant voltage and therefore at
constant current. The drain–source voltage is constant during phase I and
falls during phase II.

In region I the effective input capacitance is given by the inverse of the
slope of the voltage against charge line, i.e. $C_{eff}$ is $Q_1/7$, i.e. 1 nF. The
charging time constant $\tau$ is therefore $50 \times 1 \times 10^{-9}$, or 50 ns, and with the
gate voltage charging exponentially to 10 V, the gate voltage is given by

$$V_{GS}(t) = 10(1 - \exp(-t/\tau))$$

The time taken to charge the gate to 7 V is therefore 60 ns. In region II
the charging current is constant at $(10 - 7)/50 = 0.06$ A and the time taken
to supply the charge $Q = Q_1 - Q_2$ is given by

$$t = \frac{Q}{I} = \frac{7 \times 10^{-9}}{0.06} = 117 \times 10^{-9} \text{s}$$

Thus the delay time to start of turn-on is 60 ns and the rise time 117 ns,
giving a total turn-on time of 177 ns.

The gate and drain–source voltages, as a function of time, are illustrated
in Figure 1.10. The voltage rises exponentially during phase I and is
constant during phase II. During the third phase the voltage continues to
rise exponentially and the drain–source voltage is very low. The time delay
between the application of the gate voltage and the drain–source voltage
becoming zero is the turn-on time.

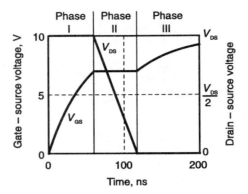

**Figure 1.10** *The gate–source and drain–source voltages (V$_{GS}$ and V$_{DS}$) of a MOSFET as it turns on*

The ideas used in Example 1.2 may be used to analyze the turn-off of the transistor. In this case the gate starts off charged to the on-state gate voltage, and then decays exponentially to the start of the flat region of the voltage against charge characteristic, when the transistor starts to turn off. The gate voltage remains constant while the drain–source voltage rises, then decays exponentially as the gate discharges fully.

## 1.5 Insulated gate bipolar transistors

### 1.5.1 Structure

A major limitation of power MOSFETs as electronic switches is that the value of $R_{DS(on)}$ increases as the voltage rating of the device is increased. This problem arises because the doping of the drain region of the device must be reduced as the voltage rating is increased, lowering its conductivity. The insulated gate bipolar transistor (IGBT) attempts to overcome this limitation by using a forward-biased *p–n* junction to inject minority carriers into this region so as to increase the

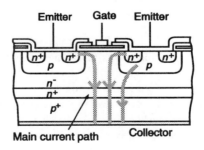

**Figure 1.11** *The cross-sectional diagram of an IGBT*

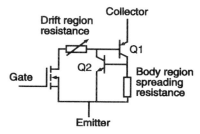

**Figure 1.12** *A circuit model for an IGBT*

conductivity. The structure of an IGBT is shown in Figure 1.11 and is essentially similar to that of a MOSFET with the addition of a $p^+$ layer. The lower $p$–$n$ junction is forward biased when the device is *on* and injects holes into the $n^-$ layer enhancing its conductivity. The main current flow is through the enhancement channel as for a MOSFET, although some of the holes will flow to the $p$ well giving bipolar transistor action.

An IGBT can be modelled using the equivalent circuit shown in Figure 1.12. The main current flow is through the MOSFET and the base of the $p$–$n$–$p$ transistor, Q1, although some of the current will also flow through the collector of Q1. The variable resistance is to emphasize that the effective resistance in series with the drain of the MOSFET is reduced by the minority carriers injected into the drain region. The parasitic $n$–$p$–$n$ transistor, Q2, has its base–emitter junction shorted out by emitter metallization. Unless this short is able to stop Q2 turning on, positive feedback to the $p$–$n$–$p$ transistor will cause the device to 'latch up', in which case, the current will no longer be controlled by the gate. This problem has been largely eliminated from the present generation of devices.

There does not seem to be universal agreement yet about the appropriate symbol for the IGBT. Two symbols that are used are shown in Figure 1.13. There also seems to be uncertainty about the nomenclature for the electrodes. Some authors use BJT type terminology of emitter and collector, others use MOSFET terminology of source and drain.

### 1.5.2 Characteristics

The gate controls the collector (or drain) current in very much the same way as it controls the drain current in a MOSFET. The main difference is that the injection of

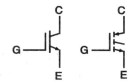

**Figure 1.13** *Symbols for an IGBT*

minority carriers reduces the on-state resistance. When used as a switch the device turns on quickly in a manner very similar to a MOSFET, but the turn-off time is an order of magnitude slower than the turn on time. The slow turn off occurs because all the injected minority carriers in the drain region must recombine or be removed before the device can block the flow of current.

Driving IGBTs is very similar to driving MOSFETs. They have large gate capacitances and again are best considered as charge controlled.

IGBTs have characteristics that combine some of the features of a bipolar transistor (low forward-voltage drop) with those of a MOSFET (high input resistance). They have found application in traction motor drives and similar circuits, where high voltage, high power, and speed of switching are all required.

## 1.6 Thyristors

Thyristors, or silicon-controlled rectifiers (SCRs), are the oldest and best-established electronic power switches. They differ from transistors in that they are latching devices which are turned on by a gate pulse, and then remain conducting until the current falls to zero. The basic thyristor will conduct only in one direction and blocks reverse voltage. Thyristors have been replaced in many low- or moderate-power circuits by other devices, which can more readily be turned off. However, they can be constructed with high voltage ratings ($>5$ kV) and very large current ratings ($>5$ kA), with an on-state voltage drop of only 1–3 V. There are a number of variations on the basic thyristor. The triac can be triggered to conduct in either direction, and is useful for controlling low-power a.c. lighting, motors, etc. It is not suitable for use at high power. The gate turn-off thyristor (GTO) is a thyristor which may be turned off by applying a (rather large) reverse current pulse through the gate. GTOs are extensively used in high-power systems because of their ability to handle voltages and currents which are beyond the range of transistors. The asymmetrical thyristor is a thyristor which is unable to withstand more than about 10 V in the reverse direction, but is able to turn off quickly (in a few microseconds). Here only the basic thyristor will be considered.

### 1.6.1  Structure and operation

The thyristor is a four-layer $n$–$p$–$n$–$p$ junction structure as shown in Figure 1.14. The easiest way to understand its operation is to consider the two-transistor equivalent circuit shown in Figure 1.15. If the anode is positive and the cathode negative there are two possible stable states. If both transistors are off then neither transistor has any base current and so the device will remain off. If one of the transistors is turned on by injecting a current between the gate and the cathode the $n$–$p$–$n$ transistor will supply the base current for the $p$–$n$–$p$ transistor, turning it on and it in turn will supply base current for the $n$–$p$–$n$ transistor. The device will therefore latch into the on-state and will remain on if the gate drive is removed.

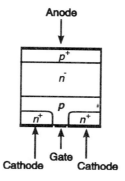

**Figure 1.14** *A simplified diagram of a thyristor*

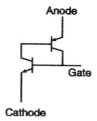

**Figure 1.15** *The two-transistor model of a thyristor*

To turn off the thyristor the current must be reduced until the effective current gain of the two transistors has fallen so far that conduction cannot be maintained and the device will start to turn off. The minimum current to maintain conduction is the *holding current*, $I_H$. Since it is a minority carrier device the stored charge must recombine before the forward voltage can be reapplied without the device returning to the conducting state. The stored charge will also result in reverse recovery transients similar to those of a junction rectifier.

### 1.6.2 Ratings

Thyristors are rated by their average forward current, $I_{T(AV)}$ and r.m.s. forward current $I_{T(RMS)}$, together with a surge rating, $I_{TM}$, usually for half a cycle duration at the power frequency. The voltage ratings are the repetitive peak off-state voltage, $V_{DRM}$, and the repetitive peak reverse voltage, $V_{RRM}$. If forward biased the thyristor may be turned on by a gate pulse, by excessive voltage (breakdown puts the device in the conducting state), or by an excessively high rate of voltage rise. There is therefore a maximum rate of voltage rise in order to avoid false triggers. There is also a maximum allowable rate of current rise when the device is turned on. Exceeding this rate of current rise may not allow uniform current distribution throughout the device, leading to local overheating and permanent damage.

Thyristors are robust devices, which frequently have surge current ratings of up to ten times their average forward current rating. To some extent they have been replaced by more recent devices like the IGBT. However, they are still the preferred devices at the highest power levels.

### 1.6.3 Driving thyristors

Unlike transistors, thyristors require only a short pulse to turn them on. Indeed it is generally undesirable to leave a continuous gate current once the device is triggered as it increases the power dissipation. In some controlled rectifier circuits the thyristor may not be in a state that it may be triggered until a time determined by the load conditions. In this case it is usual, rather than supplying a single pulse, to supply a train of pulses so that if the first pulse is unable to trigger the thyristor a later pulse will, thus avoiding a missing cycle. In most thyristor circuits it is not possible for the cathode of the thyristor to be at earth (or even a constant) potential: isolation of the drive is essential. Since only a short pulse (or train of short pulses) is required the isolation is conveniently provided by a small pulse transformer. A simple circuit that provides a train of pulses is shown in Figure 1.16. A high-frequency pulse train is gated to provide a burst of pulses when the control input is high. The transistor switches the current in the primary of the pulse transformer. The diode D1 and resistor R1 are necessary to discharge the energy stored in the primary inductance of the transformer. The problems of switching current in an inductive load will be discussed in Chapter 6. Resistor R2 provides a low-impedance path for the gate of the thyristor to help avoid false triggering when the voltage is rising rapidly across the thyristor and D2 prevents a negative voltage appearing on the thyristor gate.

## Summary

In this chapter the most widely used active power devices have been introduced. The aim has been to show the structure of these devices and indicate their most important characteristics so that their use in the circuits presented in later chapters may be understood. For transistors, the emphasis has been on their use as switches

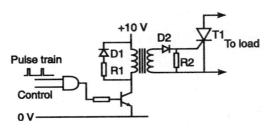

**Figure 1.16**  *A simple isolated drive circuit for a thyristor*

since this is the way in which they are usually used in power electronic circuits. Very brief consideration has also been given to how these various devices may be driven as switches.

## Tutorial questions

**1.1**   What are the relative merits and limitations of (a) silicon junction diodes and (b) Schottky diodes when used as rectifiers at low or high frequencies?

**1.2**   What is meant by the reverse recovery time of a junction diode?

**1.3**   Sketch the cross-section of a power bipolar transistor.

**1.4**   Draw a diagram showing the safe operating area of a bipolar transistor and explain the origin of each of the bounding lines.

**1.5**   What is meant by saturation in a bipolar transistor? Why is it generally avoided if high switching speeds are required?

**1.6**   Describe the operation of the Baker's clamp circuit used to avoid saturation of a bipolar switch.

**1.7**   Sketch the cross-section of a power MOSFET which uses a vertical DMOS structure. Indicate the flow of current.

**1.8**   Explain why a DMOS transistor always conducts current in the reverse direction.

**1.9**   Why does a MOSFET power switch require a low-impedance, high-current gate driver?

**1.10**   Compare the structure of an IGBT with that of a vertical MOSFET.

**1.11**   Compare the relative merits of bipolar transistors, MOSFETs and IGBTs when used as power switches.

**1.12**   Using the two-transistor model explain the operation of a thyristor.

## References

Jayant Baliga, B., *Modern Power Devices*, John Wiley, New York, 1987.

Mohan, N., Undeland, T. M. and Robbins, W. P., *Power Electronics; Converters, Applications and Design*, 2nd edition, John Wiley, New York, 1995, chapters 19–26.

## CHAPTER 2

# *Thermal management*

## 2.1 Introduction

In any power electronic circuit power will be dissipated in both the active and passive components. This produces heat which must be transferred away from the component and dispersed, usually in the air surrounding the equipment, although occasionally when the power is very large, or the volume small, liquid cooling may be preferred to transport the heat to a remote heat exchanger. For an inductor, transformer or capacitor of the size found in most moderate- or low-power electronic equipment the surface of the component is sufficient to provide adequate cooling, and special measures are not required. For a semiconductor device, however, it is most important to remove the heat at a rate high enough to limit the *junction temperature* to a reasonable value (note that the term *junction temperature* is applied to the temperature of the active region of any semiconductor device, not just bipolar junction devices). This is achieved by the use of a *heatsink* which conducts the heat away from the small active region of the device and has a large surface area which dissipates the heat into the ambient air. Heatsinks are also frequently used to cool resistors that are required to dissipate a power in excess of a few watts.

The lifetime of semiconductor devices decreases rapidly with increasing temperature and so it is essential that the manufacturer's maximum junction temperature is not exceeded. However, heatsinks are bulky and expensive and so careful thermal design is necessary to avoid excessive cost or size, while meeting the requirements for maximum temperature. The objective in this chapter is to introduce the simplified models that are used to enable the thermal calculations to be made with adequate accuracy.

## 2.2 Transport of heat

There are three mechanisms by which heat is transported:

1. Conduction, which occurs in all media, and is heat transported through the medium by the interaction of the molecules
2. Convection, which occurs in fluids and arises from the bulk movement of the heated fluid, either due to the change of density with temperature which leads to

*natural convection* or by the use of a fan or some other means of forcing the fluid to circulate which leads to *forced convection*

3. Radiation, which is the transport of heat as electromagnetic radiation and becomes more significant as the temperature rises.

Here we shall be primarily concerned with conduction, which transports the heat away from a semiconductor junction and spreads it out across the heatsink. We shall also consider convection, which determines the efficiency with which the heatsink can dissipate the heat to the surrounding air. Radiation is usually neglected, which will in most cases lead to conservative design. Note, however, that heatsink manufacturers usually paint their products black to enhance radiation, which may make a significant contribution to the heat dissipation.

### 2.2.1 Thermal conduction in the steady state

In general, thermal conduction must be considered as a problem involving three spatial directions and time, which may be expressed mathematically as a partial differential equation (the Fourier heat conduction equation). Fortunately, for most situations involving the cooling of electronic devices a much simpler approach is possible based on the use of lumped elements and thermal resistances.

Consider a bar of length $l$ and cross-section $A$. Once a steady state has been reached, the flow of heat along the bar is given by

$$P_{\text{cond}} = \frac{\lambda A \Delta T}{l} \tag{2.1}$$

where $P_{\text{cond}}$ is the thermal power conducted, $\lambda$ is the thermal conductivity and $\Delta T$ is the temperature difference between the ends of the bar. The quantity $l/\lambda A$, which depends only on the dimensions of the bar and its material, may be replaced by the *thermal resistance*, $R_\theta$. Note that the leakage of heat from the sides of the bar has been neglected. Heat flows in or out only at the ends.

The concept of thermal resistance can be applied to more complex shapes provided the flow of heat into the object all occurs at one uniform temperature and the flow of heat out takes place at a second uniform temperature. Thus for a transistor in its package the junction of the device is assumed to be at a uniform temperature, $T_j$, and the case at a temperature $T_c$ also assumed uniform, then the flow of heat out of the package is determined by the thermal resistance from junction to case $R_{\theta jc}$. The power dissipated in the transistor is treated as a discrete source at the junction temperature. The relation between power dissipated, $P_d$, and the junction and case temperatures is given by

$$P_d = \frac{(T_j - T_c)}{R_{\theta jc}} \tag{2.2}$$

### 2.2.2 Convection cooling of a heatsink

Convection is a much more complex process than conduction. However, it is observed experimentally that an object in still air at a uniform temperature, $T_s$, loses heat at a rate which is proportional to the difference between $T_s$ and the ambient temperature of the air, $T_a$ (at least to a first approximation). It is therefore possible to assign a thermal resistance $R_{\theta sa}$ to this process. The rate of loss of heat will depend on the surface area of the object as well as, in general, on the orientation. Heatsinks used to cool electronic devices are usually fabricated from extruded aluminium, with fins to increase the surface area. The heatsink will have its lowest thermal resistance if it is mounted so that the grooves between the fins are vertical and air is free to flow up the grooves as it heats and expands. Since the thermal conductivity of aluminium is low the thermal resistance across the heatsink is usually much less than the thermal resistance from the heatsink to the ambient air. Hence the assumption that the heatsink is at uniform temperature will be valid.

Where the area of heatsink required is excessive or the free circulation of air may be restricted then a fan may be used to induce the flow of air across the heatsink (forced convection). This may greatly reduce the thermal resistance, and for optimum performance the heatsink should have more closely spaced fins than for natural convection.

### 2.2.3 Radiative cooling of a heatsink

The rate at which heat is transferred by radiation from a surface is given by the Stefan–Boltzmann law,

$$P_{rad} = \sigma \alpha A (T_s^4 - T_a^4) \tag{2.3}$$

where $\sigma$ is the Stefan–Boltzmann constant ($5.67 \times 10^{-8}$ J s$^{-1}$ m$^{-2}$ deg$^{-4}$), $\alpha$ the emissivity of the surface, $A$ the area, $T_S$ the surface temperature (in K) and $T_a$ the ambient temperature. The emissivity, $\alpha$, varies from less than 0.1 for a polished surface to about 0.9 for a matt-black one (e.g. black anodized aluminium). Obviously, the heat loss is not proportional to the temperature difference, so if an attempt is made to define the thermal resistance for radiation the value obtained will vary with the ambient temperature and the temperature difference. For a typical heatsink with a matt-black finish radiation may account for about 30% of the total heat dissipation at a heatsink temperature of 100°C and an ambient temperature of 20°C. With a smaller temperature difference, radiation will be less significant.

### 2.2.4 Thermal contact resistance

When the surfaces of two thermal conductors are in contact there will be a thermal resistance between the surfaces which will depend on the surface area nominally in contact and upon the flatness of the surfaces. Microscopic roughness and distorted surfaces may lead to thermal resistances that are very much higher than expected. A common method of reducing the contact resistance is to use heatsink compound, a paste comprising metal oxides in a base of silicone, or other grease. A very thin film of the paste across the interface fills the voids and improves thermal contact. An excess of paste will result in poorer thermal contact. As an example of the improvement to be expected, a TO-3 transistor package mounted on a flat metal heatsink might have a contact resistance of $0.5°C\ W^{-1}$ without a compound and $0.1°C\ W^{-1}$ with a compound. If the transistor is electrically insulated from the heatsink with a 0.08 mm thick mica washer then the thermal resistance would be about $1.3°C\ W^{-1}$ without a compound decreasing to about $0.4°C\ W^{-1}$ with correctly applied heatsink compound.

## 2.3 The thermal circuit model

Since both conduction and convection cooling of the heatsink can be modelled by thermal resistances, it is possible to model the whole thermal dissipation process by an equivalent electrical circuit, provided radiation is neglected. Power is treated as analogous to current, temperature as analogous to voltage and thermal resistance as analogous to electrical resistance. If we consider a typical situation of a power transistor mounted on a heatsink using a washer for insulation (Figure 2.1(a)) then this translates to the electrical circuit of Figure 2.1(b). The reference node is the ambient temperature and the power dissipation in the transistor, which is represented by the current source $P_d$, is returned to the reference node. Thermal resistances in series or parallel behave just like electrical resistances and the usual formulae apply. The current in any branch of the circuit is the local power flow, while the node voltages are the temperatures of the junctions between the various components.

### EXAMPLE 2.1

Two identical transistors Q1 and Q2 each with a thermal resistance

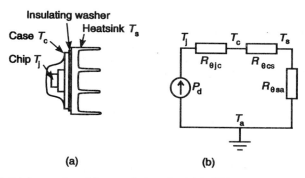

(a)                              (b)

**Figure 2.1** (a) A power transistor mounted on a heatsink; (b) its equivalent circuit model

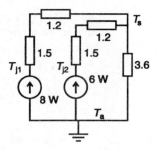

**Figure 2.2** *Equivalent circuit model for Example 2.1*

between the junction and the case $R_{\theta jc}$ of 1.5 W °C$^{-1}$, are mounted on a common heatsink using washers which give an effective thermal resistance of 1.2 W °C$^{-1}$ between case and heatsink, $R_{\theta cs}$. The heatsink has a thermal resistance of 3.6W °C$^{-1}$. If Q1 dissipates a power of 8 W and Q2 a power of 6 W, what are their junction temperatures if the ambient temperature is 40°C?

**SOLUTION**

The equivalent circuit is shown in Figure 2.2. Power from both transistors flows through the heatsink, hence the heatsink temperature is given by

$$T_s = T_a + (8 + 6) \times 3.6 = 90.4°C$$

Adding $R_{\theta jc}$ to $R_{\theta cs}$ gives the junction-to-sink resistance as 2.7 W °C$^{-1}$. Hence the temperatures of the two junctions are given by

$$T_{j1} = T_s + 8 \times 2.7 = 112°C$$
$$T_{j2} = T_s + 6 \times 2.7 = 106.2°C$$

## 2.4  Transient effects

### 2.4.1  The lumped-component thermal model

The thermal circuit model may be extended to handle transient effects by using the concept of thermal capacity. A component of the thermal circuit heats up and stores energy in a way analogous to how a capacitor stores charge. Thus we have not only thermal resistances but also thermal capacity. The thermal capacity is equal to the mass of the component multiplied by the specific heat of the material from which it is made. The concept of the thermal capacity can only be used if the whole of the component is at the same temperature. If there are significant temperature differences across the component then it is necessary to split it into several smaller elements.

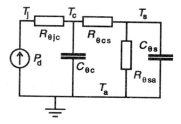

**Figure 2.3** *Equivalent circuit model including thermal capacity*

Consider a TO-3 metal-cased transistor mounted on a heatsink using an insulating washer. The heat is generated at the junction of the transistor and transported through the header to the case. Most of the thermal resistance will be located in and close to the silicon die while most of the thermal capacity of the transistor will be located in the case. Thus an approximate thermal model may be created, as shown in Figure 2.3. The transistor case has a thermal capacity $C_{\theta c}$, and the heatsink a thermal capacity $C_{\theta s}$, while the junction-to-case thermal resistance is $R_{\theta jc}$, the case-to-heatsink resistance $R_{\theta cs}$, and the thermal resistance of the heatsink is $R_{\theta sa}$. The transient response to changes in the power dissipated, $P_d$, can readily be calculated as for an electrical circuit.

### EXAMPLE 2.2

A transistor is mounted directly on an aluminium heatsink with a thermal resistance of 2°C W$^{-1}$ and a mass of 0.15 kg. The transistor is turned on at time $t = 0$ and dissipates 15 W. If the heatsink is initially at the ambient temperature of 25°C, how long will it take to reach a temperature of 40°C? The specific heat of aluminium is 900 J kg$^{-1}$ °C$^{-1}$.

### SOLUTION

The thermal capacity of the heatsink is 0.15 × 900, or 135 J °C$^{-1}$, and the equivalent thermal circuit is shown in Figure 2.4. The thermal capacity of the transistor case has been ignored. It is not necessary to know $R_{\theta jc}$ to deduce the heatsink temperature.

Writing down the equation for the temperature of the heatsink gives

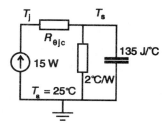

**Figure 2.4** *Equivalent circuit for Example 2.2*

$$135\frac{dT_s}{dt} + \frac{T_s - 25}{2} = 15$$

If the power is turned on at $t = 0$, then at this time $T_s$ has a value of 25, and the solution has the form

$$T_s = 25 + 30(1 - \exp[-t/270])$$

The heatsink heats up exponentially with a time constant of 270 s to a final temperature of 55°C. Setting $T_s$ to 40, and solving for $t$ gives the time for the heatsink temperature to reach 40°C as 187 s.

### 2.4.2 Transient thermal impedance

The lumped model using thermal resistances and thermal capacitance works well only where it is possible to divide the thermal path into a series of elements each at a uniform temperature, separated by thermal resistances. In order to model a semiconductor device in its package the thermal path must be split up into a number of discrete elements, each with its own thermal capacity, and separated by thermal resistances, thus forming an RC ladder network. A very simple model might involve three elements, the silicon die, the header, and the case, and is illustrated in Figure 2.5. The thermal capacity of the die will be small and that of the case large, while the thermal resistance between die and header will be larger than that between header and case. There is therefore a range of time constants involved.

To avoid the complexity introduced by this approach, the concept of transient thermal impedance is introduced. The response of the junction temperature to a step change in power is considered. At $t = 0$ the power rises from 0 to $P_d$. Then, assuming the case temperature remains constant at $T_c$, the junction will reach a temperature $T_j$ at time $t$, where

$$T_j = P_d Z_{\theta jc}(t) + T_c \tag{2.4}$$

The time dependent parameter, $Z_{\theta jc}(t)$, is the thermal impedance. Clearly, as $t$ becomes large, equilibrium is reached and $Z_{\theta jc}$ equals $R_{\theta jc}$. For brief times heat is stored in the internal thermal capacity and $Z_{\theta jc}$ will be less than $R_{\theta jc}$. If rather than

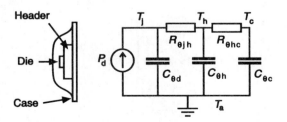

**Figure 2.5** *Equivalent circuit for thermal transport in a transistor, showing thermal capacities*

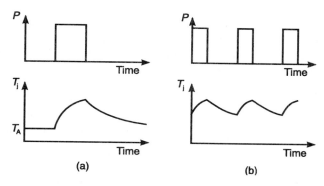

**Figure 2.6** *The variation of junction temperature ($T_j$) with time for (a) a single pulse of power and (b) repetitive pulses*

applying a step of power a pulse of duration $t$ is applied the temperature of the junction will rise until the time $t$, when it will start to fall, and the maximum temperature will be given by equation (2.4). The variation of junction temperature with time for a single pulse of dissipated power is shown in Figure 2.6(a)

In practice it is common in electronic circuits for the power to be dissipated as a train of pulses with a constant period, pulse size and duration. The temperature of the junction will vary as illustrated in Figure 2.6(b). In order to handle this circumstance the concept of thermal impedance is extended. It is assumed that the maximum junction temperature is given by equation (2.4) but, that the value of $Z_{\theta jc}(t)$ now also depends on the duty ratio of the pulses. The duty ratio, $D$, is defined as the ratio of the pulse duration to the period. In this case the parameter $Z_{\theta jc}(t)$ is no longer the thermal impedance, although it is closely related to it. For repetitive pulses $Z_{\theta jc}(t)$ is often referred to as the *effective thermal impedance*. The data for $Z_{\theta jc}(t)$ are usually presented graphically as in Figure 2.7, where, as is usual in data sheets, the thermal impedance $Z_{\theta jc}(t)$ has been normalized by dividing by the thermal resistance $R_{\theta jc}$.

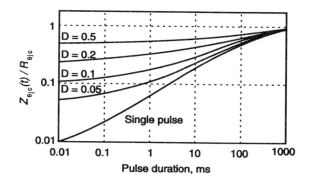

**Figure 2.7** *The effective thermal impedance*

The repetition period of the pulses will usually be sufficiently short compared with the thermal time constant of the case of the transistor (or other device) that the case temperature and the heatsink temperature may be considered constant. In this case the mean power dissipation will determine the case temperature, and, assuming a steady state has been reached, the temperatures external to the device may be calculated using mean power and thermal resistances.

If the power pulses are not of rectangular shape, an equivalent rectangular pulse of the same peak power and total energy may be used to obtain an approximate solution.

### EXAMPLE 2.3

The power dissipated in a transistor may be represented by a train of pulses of duration $100\,\mu s$ with a duty ratio of 0.2 and a peak power of 30 W. The transistor is mounted directly on a heatsink with a thermal resistance of 5 W °C$^{-1}$. If the effective thermal impedance of the transistor is 0.4°C W$^{-1}$ for 0.1 ms pulses with a duty ratio of 0.2, what is the peak junction temperature if the ambient temperature is 40°C?

### SOLUTION

The peak junction temperature is given by

$$T_{j\text{max}} = 0.4 \times 30 + T_c = 12 + T_c$$

It is assumed that the case and heatsink have a large thermal time constant, so that the case temperature is constant. The case temperature is then found by taking the average power dissipated and the thermal resistance of the heatsink:

$$T_c = 30 \times 0.2 \times 5 + 40 = 70° C$$

Hence the maximum junction temperature is given by

$$T_{j\text{max}} = 70 + 12 = 82° C$$

## Summary

In this chapter the basic techniques used to perform thermal calculations in electronic equipment have been described. The use of these simplified one-dimensional models based on an electrical circuit analogy is usually adequate for most purposes. The necessary thermal resistances and thermal impedances are specified by the manufacturers of power transistors and other active devices, as are the thermal resistances for proprietary heatsinks.

# Self-assessment questions

2.1  A mica washer 0.08 mm in thickness is used to insulate a transistor in a TO-3 package from a heatsink. If the effective area of the contact surface of the package is 400 mm$^2$, what is the thermal resistance between the package and the heatsink, neglecting the contact resistance? The thermal conductivity of mica is 0.6 W m$^{-1}$ °C$^{-1}$.

2.2  A rectifier diode has two copper leads each 1 mm in diameter. The silicon die is sandwiched between the leads and makes good thermal contact with them. The leads of the rectifier are connected to two terminals which are at a constant temperature of 30°C. The length of lead from each terminal to the die is 10 mm. Neglecting heat loss by convection to the surrounding air, estimate the steady-state junction temperature if the power dissipated in the diode is 2.5 W. Assume that the thermal conductivity of copper is 380 W m$^{-1}$ °C$^{-1}$.

2.3  A transistor is mounted on a heatsink using a mica washer for insulation. The resistance from junction to case, $R_{\theta jc}$, of the transistor is 1.5°C W$^{-1}$ and the thermal resistance of the washer, including contact resistance, is $R_{\theta cs}$, is 1°C W$^{-1}$. If the power dissipated is 20 W, the ambient temperature is 40°C and the junction temperature must not exceed 125°C, what is the maximum thermal resistance of the heatsink?

2.4  Two identical transistors are mounted on a common heatsink, which has a thermal resistance of 2°C W$^{-1}$. If the thermal resistance of each transistor from junction to case, $R_{\theta jc}$, is 2°C W$^{-1}$ and the contact resistance between transistor and heatsink is 0.2°C W$^{-1}$, what is the maximum power that may be dissipated (a) with all the power dissipated in one transistor and (b) with the power divided equally between the two transistors? Assume an ambient temperature of 55°C and a maximum junction temperature of 125°C.

2.5  A transistor in a metal package is mounted in free air (no heatsink) at an ambient temperature of 30°C. The package has a thermal resistance from junction to case, $R_{\theta jc}$, of 6°C W$^{-1}$, a thermal resistance from case to ambient air of 60°C W$^{-1}$ and a thermal capacity for the case of 0.25 J °C$^{-1}$ s$^{-1}$. The thermal capacity may be assumed to be lumped at the outside of the case, while $R_{\theta jc}$ is between case and junction. The power dissipated in the transistor is 2.5 W and is switched on at time $t = 0$. How long will it take for the junction temperature to reach 125°C, starting from ambient?

2.6  A transistor dissipates power in pulses with an amplitude of 100 W, a duration of 1 ms and a period of 10 ms. The value of the effective thermal impedance from junction to case, $Z_{\theta jc}$, for pulses of this duration and duty ratio is 0.18°C W$^{-1}$. What thermal resistance is required for the heatsink if the junction temperature is not to exceed 125°C when the ambient temperature is 40°C? Neglect the thermal resistance between the transistor case and the heatsink.

## Tutorial questions

**2.1**  Define thermal resistance.

**2.2**  Why is the use of thermal resistance appropriate when heat is transported by conduction and convection, but not radiation?

**2.3**  Why does the orientation of a heatsink affect its thermal resistance?

**2.4**  What is the cause of thermal contact resistance, and how may a heatsink compound be used to reduce it?

**2.5**  In what circumstances is it useful to use thermal capacity and thermal resistance to deduce the temporal response of a thermal system?

## Reference

Mohan, N., Undeland, T. M. and Robbins, W. P., *Power Electronics; Converters, Applications and Design*, 2nd edition, John Wiley, New York, 1995, Chapters 19–26.

## CHAPTER 3

# Magnetic components

## 3.1 Introduction

In low-power electronics, transformers and inductors are usually avoided. These components cannot be integrated on silicon (except in very small values) and are frequently far from ideal in their characteristics. Modern circuit design is generally able to avoid transformers or inductors in the signal path, except in a few special cases. In power electronics, however, transformers and inductors are frequently essential. Transformers are used not only because they are an efficient component for changing the voltage in a.c. power systems but also because they are able to provide a high degree of electrical isolation between the primary and secondary circuits. Inductors are used in several ways, the two most common being as efficient energy storage and transfer elements and as essential components in low-loss filters. D.C.-to-d.c. power conversion circuits frequently make use of the ability of an inductor to store energy at one voltage and deliver it at a different one (higher or lower), and to do so without loss (for an ideal inductor).

In general, it is not possible to purchase suitable transformers or inductors off the shelf as the requirements are so diverse. While there are limited ranges of standard components available, it is frequently necessary to have inductors or transformers designed for the particular application. The designer of power electronic equipment needs to have an understanding of the methods of construction and the characteristics of transformers and inductors to be able to specify the components that will be required and obtain optimum performance.

In this chapter some familiarity with basic electromagnetism is assumed. The aim of the chapter is to revise some of the basic theory needed to understand the behaviour of inductors and transformers.

## 3.2 Magnetic circuits

The inductors and transformers used in power electronics are generally made using ferromagnetic cores with a large relative permeability. For such cores it is possible to use the magnetic circuit model to greatly simplify the calculation of inductance. This model is based upon three simplifying assumptions:

1. All the magnetic flux is confined within the core.

2.  The magnetic flux density across any core section perpendicular to the flux lines is uniform.
3.  The flux density $B$ is proportional to the magnetic field strength $H$.

For assumption (1) to be valid the relative permeability of the core, $\mu_r$, must be large so that the field lines are concentrated within the core. This assumption is usually reasonable for ferromagnetic cores of iron or high-permeability ferrite where $\mu_r > 1000$. Assumption (2) will not be well satisfied near sharp bends in the core, but usually the errors introduced can be tolerated. This assumption is used to relate the total flux $\Phi$ to the flux density $B$, i.e.

$$\Phi = BA$$

where $A$ is the cross-sectional area of the core. The final assumption (3) will be considered in detail in Section 3.3.

The starting point for the magnetic circuit model is Ampère's law;

$$\oint H.dl = \Sigma I \tag{3.1}$$

where the line integral is round a closed loop and $\Sigma I$ is the sum of all the electric currents passing through the loop. Consider a simple core with a constant cross-sectional area as shown in Figure 3.1. A coil of $N$ turns is wound around the core and carries a current $I$. Taking any closed path around the core then Ampère's law gives

$$\oint H.dl = NI \tag{3.2}$$

The quantity $NI$ is called the magnetomotive force (m.m.f.), since it drives the flux around the magnetic path. If a path is chosen which passes through the centre of each cross-section of the core, then $H$ will be approximately parallel to the path at all points and the integral may be written

$$\oint H.dl = Hl_m \tag{3.3}$$

where $l_m$ is the length of this path, and is referred to as the magnetic path length. Using assumptions (2) and (3) $H$ may be replaced by

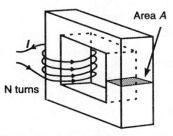

**Figure 3.1** *Inductor wound on a simple ferromagnetic core*

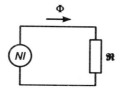

**Figure 3.2** *Magnetic circuit model for the core shown in Figure 3.1*

$$H = \frac{B}{\mu_r \mu_0} = \frac{\Phi}{\mu_r \mu_0 A} \tag{3.4}$$

Using equations (3.2)–(3.4) an expression relating the flux to the magnetomotive force is obtained:

$$\frac{\Phi l_m}{\mu_r \mu_0 A} = \Phi \Re = NI \tag{3.5}$$

The quantity $\Re = l_m/\mu_r \mu_0 A$ is the *reluctance* of the core. An analogy is usually drawn between equation (3.5) and Ohm's law, with the flux $\Phi$ replacing the current, the reluctance $\Re$ replacing the resistance and the magnetomotive force $NI$ replacing the e.m.f.

An electric circuit model to describe the simple magnetic system of Figure 3.1 is shown in Figure 3.2. If the core is more complex than that shown in Figure 3.1, with sections of different cross-sectional area, then the core is considered as a series of elements each with their own cross-sectional area, $A_i$, and magnetic path length, $l_i$. The reluctance $\Re_i$ of each element is obtained from the equation

$$\Re_i = \frac{l_i}{\mu_r \mu_0 A_i} \tag{3.6}$$

If the flux passes through the core sections in series then the reluctances add like resistances in series. If core sections are in parallel their reluctances add like resistances in parallel. The use of the magnetic circuit model for analysing more complex core shapes is best illustrated by a specific example.

### EXAMPLE 3.1

An E-I core of the type widely used for single-phase transformers is shown in Figure 3.3. Calculate the reluctance of the core for an m.m.f. produced by a coil wound around the central pole of the core. The core dimensions are shown in the diagram and the relative permeability $\mu_r$ is 1000.

### SOLUTION

An appropriate equivalent magnetic circuit is shown in Figure 3.4(a). The m.m.f. is in series with the central pole of the core with reluctance $\Re_p$ and drives a flux $\Phi$, which is split equally between the two side limbs of reluctance $\Re_l$. The magnetic path is shown in Figure 3.4(b), for the centre

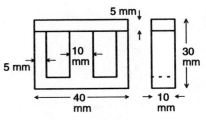

**Figure 3.3** *E–I transformer core*

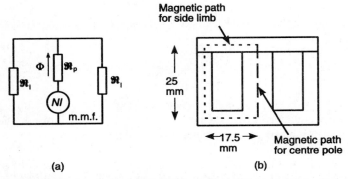

**Figure 3.4** *(a) Magnetic circuit model for E–I core; (b) magnetic paths for E–I core*

pole and one of the limbs. It can be seen that the pole has a magnetic path length of 25 mm, while the limb has a path length of $17.5 + 25 + 17.5 = 60$ mm. Hence the reluctances may readily be calculated, and are given by

$$\Re_p = \frac{l}{\mu_r \mu_0 A}$$

$$= \frac{25 \times 10^{-3}}{1000 \times 4\pi \times 10^{-7} \times 0.01^2}$$

$$= 2 \times 10^5 H^{-1}$$

and

$$\Re_l = \frac{60 \times 10^{-3}}{1000 \times 4\pi \times 10^{-7} \times 0.01 \times 0.005}$$

$$= 9.55 \times 10^5 H^{-1}$$

The two side-limbs are in parallel and have the same reluctance, hence the total reluctance of the core is given by

$$\Re = \Re_p + 0.5\Re_l = 6.78 \times 10^5 H^{-1}$$

The reluctance is a useful parameter for a core, since, as will be shown in

Section 3.4, it relates the inductance of a coil of wire wound around the core to the number of turns in the coil.

## 3.3 Properties of ferromagnetic materials and cores

Ferromagnetic materials show very complex magnetic characteristics in that they are non-linear, they display hysteresis and their properties change with temperature. All these characteristics are highly undesirable for the designer of inductors or transformers. There are two groups of ferromagnetic material widely used in magnetic cores for power electronic applications: metallic alloys, usually based on iron, and ferrites, which are ceramic materials. Laminated metallic cores are in general use at power-line frequencies and up to about 1 kHz. At higher frequencies ferrite materials dominate up to a frequency of at least 1 MHz. Metallic dust cores are useful up to the low MHz region, but generally have a rather low permeability.

### 3.3.1 Hysteresis and permeability

All ferromagnetic materials show hysteresis, i.e. the present value of the flux density, $B$, depends not only on the present magnetic field strength, $H$, but also on the magnetic history. A typical symmetric hysteresis curve is shown in Figure 3.5. This type of curve is obtained by applying an alternating, sinusoidal, magnetic field $H$, and measuring the flux density $B$. The intersection of the curve with the $H = 0$ axis is the remanence $B_r$ and the intercept with the $B = 0$ axis is the coercivity $H_c$. The remanence is the residual flux density remaining when the magnetizing field, or magnetomotive force, is removed. The coercivity is the magnetizing field ($H$) that must be applied in order to reduce the residual magnetic flux density to zero. These two parameters are a measure of how good the magnetic material is as a permanent magnet. To make a good permanent magnet the remanence must be large to make a strong magnet, and the coercivity must be large so that the magnet is not easily demagnetized by external magnetic fields. Such a material is called a *hard* magnetic material. For a *soft* magnetic material, such as is used in transformer

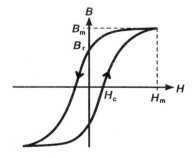

**Figure 3.5** *Symmetric hysteresis loop*

or inductor cores, $B_r$ and $H_c$ should be as small as possible.

The relative permeability of a ferromagnetic material is defined as the ratio of the maximum values of $B$ and $H$ for a symmetric hysteresis curve divided by the permeability of free space, i.e.

$$\mu_r = \frac{B_m}{\mu_0 H_m} \tag{3.7}$$

where $B_m$ and $H_m$ are as shown in Figure 3.5. The initial value for the relative permeability, $\mu_i$, is the value with small $H_m$ and is generally quite small. It increases with increasing $H_m$ to a maximum, often of around three times the initial value and then decreases as the material approaches saturation.

This definition of permeability is appropriate where the inductor or transformer carries a.c. For an inductor carrying d.c. with only a small a.c. component a more appropriate definition is

$$\mu_{inc} = \frac{\Delta B}{\mu_0 \Delta H} \tag{3.8}$$

where $\Delta B$ and $\Delta H$ are the peak-to-peak amplitudes of the a.c. modulation of the mean fields and $\mu_{inc}$ is the incremental permeability. As the magnetic field is cycled around its mean value the locus of $B$ and $H$ will trace out a minor hysteresis loop. The value of $\mu_{inc}$ is generally less than $\mu_r$ and decreases with increasing magnetic field strength.

Hysteresis leads to power dissipation as the flux is cycled. The stored energy density in a magnetic field is given by

$$E(B) = \int_0^B H \, \mathrm{d}B \tag{3.9}$$

When applied to a hysteresis loop the energy supplied is greater than the energy recovered as the loop is traversed, and the energy loss for one cycle is equal to the area enclosed by the loop. Hysteresis losses increase in proportion to the frequency. An empirical relation is often assumed, which gives the hysteresis loss per unit volume at a frequency, $f$, and a maximum flux density, $B_m$ (assumed sinusoidal) as

$$P = K_h B_m^z f \ \mathrm{Wm}^{-3} \tag{3.10}$$

where $K_h$ and $z$ are empirical constants. For ferrite materials hysteresis loss is the dominant loss mechanism, and manufacturers usually provide graphs showing magnetic loss per unit volume as a function of frequency for a range of maximum flux values.

### 3.3.2  Eddy currents and skin depth

Ferrite materials have a very high resistivity, hence an alternating magnetic flux does not introduce any circulating electric current. For metallic cores the resistivity is low and an alternating magnetic flux will induce large eddy currents which will dissipate power. The circulating currents will also produce a magnetic field which will oppose the applied field, limiting the penetration of the field into the core. The penetration decays with a characteristic distance called the *skin depth*, $\delta$, given by

$$\delta = \sqrt{\frac{2\rho}{\mu_r\mu_0\omega}} \qquad (3.11)$$

where $\rho$ is the resistivity and $\omega$ is the angular frequency. For a typical transformer core steel $\delta$ is of the order of 0.5 mm at 50 Hz. To reduce these problems metallic cores are used in the form of thin *laminations* insulated from each other, and stacked to give the required core dimensions. The laminations are arranged so that the field lines lie in the plane of the laminations. This forces the current eddy currents to flow in a long narrow paths as shown in Figure 3.6, reducing the magnitude of the current and its associated loss. The thickness of the laminations should, ideally, not exceed the skin depth.

The induced e.m.f. driving the eddy currents is proportional to the rate of change of flux, and hence the frequency. The power loss is proportional to the square of the e.m.f. and hence to the square of the frequency. By the use of very thin laminations and high-resistivity steel, laminated cores may be used up to about 10 kHz, but at that frequency, and higher, ferrites with their very high resistivity are usually preferred.

## 3.4 Self-inductance

For a coil of $N$ turns of wire around a ferromagnetic core of reluctance $\mathfrak{R}$, and carrying a current $i$, the flux, $\Phi$, calculated from the magnetic circuit model (equation (3.5)) is given by

$$\Phi = Ni/\mathfrak{R} \qquad (3.12)$$

Using Faraday's law of induction, the e.m.f. induced in the coil will be equal to the rate of change of the linked flux. Since, for an ideal magnetic core, all the flux will

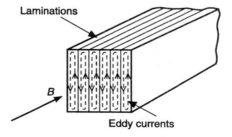

**Figure 3.6** *Eddy currents in laminated core*

be linked with each turn, the e.m.f. is given by

$$e = \frac{\mathrm{d}}{\mathrm{d}t}(N\Phi) = N\frac{\mathrm{d}\Phi}{\mathrm{d}t} = \frac{N^2}{\Re}\frac{\mathrm{d}i}{\mathrm{d}t} \qquad (3.13)$$

The relation between the applied voltage and the current for an inductor is

$$v = L\frac{\mathrm{d}i}{\mathrm{d}t} \qquad (3.14)$$

Comparing equations (3.13) and (3.14), it can be seen that the inductance, $L$ is given by

$$L = \frac{N^2}{\Re} \qquad (3.15)$$

Also using equations (3.12) and (3.15), the flux $\Phi$ and the current $i$ are linked by

$$Li = N\Phi \qquad (3.16)$$

It can be seen from equation (3.15) that the inductance is inversely proportional to $\Re$ and hence is proportional to $\mu_r$. As discussed in Section 3.3.1, the relative permeability is not well defined for ferromagnetic materials, it varies with flux density in a complex manner. Inductances wound on ferromagnetic cores may therefore be expected to be very non-linear and lossy. Fortunately, there is a simple solution, which is to insert an *air gap* in the magnetic circuit of the core.

## 3.5  The use of air gaps

### 3.5.1  Air gaps and linearity

If, rather than having a continuous high permeability core, a narrow gap (of air or other low-permeability insulating material) is introduced, then the performance of an inductor with a ferromagnetic core may be greatly improved. Consider the simple core shown in Figure 3.7. This has a cross-sectional area $A$ and a magnetic path length in the core of $l$, with a narrow air gap of length $g$.

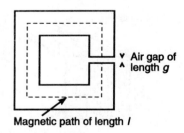

Figure 3.7  *Ferromagnetic core with air gap*

If the gap is very short compared with the linear dimensions of the core cross-section the flux density in the gap will be the same as that in the core and all the flux will cross the gap, with no flux leakage. In this case the total reluctance of the magnetic circuit is given by

$$\Re = \Re_c + \Re_g = \frac{l}{\mu_r \mu_0 A} + \frac{g}{\mu_0 A} \tag{3.17}$$

where $\Re_c$ is the reluctance of the core and $\Re_g$ the reluctance of the gap (where the relative permeability is 1). $\Re_c$ is highly non-linear, but $\Re_g$, which does not involve any ferromagnetic material is linear. If $\Re_g$ can be made much larger than $\Re_c$, then the non-linearities of $\Re_c$ will be swamped, and a more linear inductor will result.

Consider a coil of $N$ turns wound around the core of Figure 3.7. Then from equation (3.15)

$$L = \frac{N^2}{\Re_g + \Re_c} \tag{3.18}$$

Clearly if $\Re_c \ll \Re_g$ the inductor will be almost linear, but its inductance will be much less than it would have been if the core had no gap.

The size of the air gap that can be used is limited by the requirement that all the flux crosses the gap. At the gap the field will spread out, and if the width of the gap is large, field lines will escape from the core. This is referred to as *flux leakage*.

## EXAMPLE 3.2

The core described in Example 3.1 is made up from an E and an I section joined together. Instead of making a close joint a non-magnetic spacer 0.1 mm thick is inserted between the sections to make a core as illustrated in Figure 3.8. A coil of 100 turns is wound around the centre pole. Calculate the inductance of this winding.

## SOLUTION

The equivalent magnetic circuit is shown in Figure 3.9. The values of $\Re_l$ and $\Re_p$ are $9.55 \times 10^5$ H$^{-1}$ and $2 \times 10^5$ H$^{-1}$, as in Example 3.1, while the reluctance of the air gaps in the pole and the side limbs, respectively, are

Non-magnetic spacer

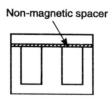

**Figure 3.8** *E–I core with non-magnetic spacer*

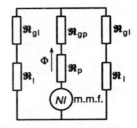

**Figure 3.9** *Magnetic circuit model for E–E core with spacer*

$$\Re_{gp} = \frac{10^{-4}}{\mu_0 \times (0.01)^2} = 7.96 \times 10^5 \text{H}^{-1}$$

and,

$$\Re_{gl} = \frac{10^{-4}}{\mu_0 \times 0.01 \times 0.005} = 1.592 \times 10^6 \text{H}^{-1}$$

The total core reluctance is given by

$$\Re = \Re_{gp} + \Re_p + 0.5(\Re_{gl} + \Re_l)$$
$$= 2.27 \times 10^6 \text{H}^{-1}$$

Hence the inductance is given by

$$L = \frac{N^2}{\Re} = \frac{100^2}{2.27 \times 10^6} = 4.40 \times 10^{-3} \text{H}$$

### 3.5.2 Air gaps and stored energy

The energy stored in a magnetic field is given by equation (3.9). If the simplifying assumption is made that $\mu_r$ is constant then the equation may be integrated to give for the stored energy density

$$\oint H \mathrm{d}B = 0.5BH = 0.5B^2/\mu_r\mu_0 \tag{3.19}$$

If the core has a maximum working flux of $B_m$ and a volume $V$, then the maximum stored energy in the core is given by

$$E_{max} = 0.5VB_m^2/\mu_r\mu_0 \tag{3.20}$$

However, not all the energy is recovered if the m.m.f. is allowed to fall to zero, as the flux density will only fall to $B_r$, the remanence. The useful stored energy is therefore

$$E = 0.5V(B_m^2 - B_r^2)/\mu_r\mu_0 \tag{3.21}$$

which may be substantially less than the maximum stored energy in equation (3.20).

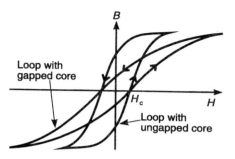

**Figure 3.10** *Effect of air gap on hysteresis loop*

The use of an air gap can substantially increase the ability to store and recover energy. If the total volume of the gap is $V_g$ then the energy stored in the gap is given by

$$E_g = 0.5 V_g B_m^2 / \mu_0 \tag{3.22}$$

Although $V_g$ may be much less than the core volume $V$, the energy density is greater by a factor $\mu_r$. Furthermore, since the gap does not show hysteresis, all the energy stored in the gap may be recovered.

If the $B$–$H$ curve is measured for a coil with an air gap, it is tilted over with respect to the $B$–$H$ curve for the ungapped core (Figure 3.10). The coercivity is unchanged, but the remanence is reduced. The total area enclosed by the loop is unchanged, since the energy lost in cycling the flux in the core is unchanged.

## EXAMPLE 3.3

For the inductor described in Example 3.2, calculate the maximum stored energy and the maximum working current if the maximum working flux density is $0.3T$.

## SOLUTION

There are two ways in which this problem may be approached. The first is to use equation (3.16):

$$Li = N\Phi$$

The maximum flux is the maximum flux density ($0.3T$) multiplied by the core area, $10^{-4}$ m$^2$, hence the maximum flux is $\Phi_{max} = 3 \times 10^{-5}$ Wb, and the maximum current is given by

$$i_{max} = \frac{100 \times 3 \times 10^{-5}}{4.4 \times 10^{-3}} = 0.68 \text{ A}$$

while the maximum stored energy is given by

$$E = 0.5Li_{max}^2 = 0.5 \times 4.4 \times 10^{-3} \times 0.68^2 = 1.02 \text{ mJ}$$

The second approach, which yields more information about where the energy is stored, is to consider the volume of the core and the energy density. The volume of the core, calculated from the dimensions in Figure 3.3, is $8 \times 10^{-6}$ m$^3$, while that of the air gap is $2 \times 10^{-8}$ m$^3$. The corresponding values for the stored energy in the core and the gap are given by

$$E_c = \frac{B^2 V_c}{2\mu_r\mu_0} = \frac{0.3^2 \times 8 \times 10^{-6}}{2 \times 1000 \times 4\pi \times 10^{-7}} = 2.86 \times 10^{-4} \text{ J}$$

$$E_g = \frac{B^2 V_g}{2\mu_0} = \frac{0.3^2 \times 2 \times 10^{-8}}{2 \times 4\pi \times 10^{-7}} = 7.16 \times 10^{-4} \text{ J}$$

The total stored energy is therefore 1.002 mJ.

From the maximum stored energy the maximum working current is easily found:

$$i_{max} = \sqrt{\frac{2E_{max}}{L}} = \sqrt{\frac{2 \times 1.002 \times 10^{-3}}{4.4 \times 10^{-3}}} = 0.675 \text{ A}$$

The two methods give answers which differ slightly. This is because the magnetic circuit theory makes an assumption about the uniformity of the field density that is only an approximation, and the method of estimating the magnetic path length is crude. The two methods make different assumptions, and exact agreement should not be expected. The second method shows that in this example about 70% of the energy is stored in the air gap.

## 3.6 Transformers

There are five main groups of transformers used in power electronics:

1. Power transformers for power line frequency (50 or 60 Hz)
2. High-frequency power transformers (20 kHz to 1 MHz)
3. Pulse transformers for isolation of control pulses to power switches
4. Current transformers for measuring pulsed currents in semiconductor switches
5. Small-signal isolation transformers for use in control loops.

The first three types are the most important and most widely used. Power transformers for 50 or 60 Hz operation always use laminated steel cores, while high-frequency power transformers use ferrite ones. Pulse transformers may use either, depending on application and frequency. Transformer cores are usually constructed with no air gap to minimize reluctance and flux leakage.

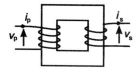

**Figure 3.11** *Simple transformer*

A simple transformer has two windings on a common core (Figure 3.11). The same flux is linked to both windings. Neglecting the resistance of the windings, the primary voltage,$v_p$, and the secondary voltage, $v_s$, are given by

$$v_p = N_p \frac{d\Phi}{dt}$$

$$v_s = N_s \frac{d\Phi}{dt}$$

(3.23)

where $N_p$ is the number of turns on the primary winding and $N_s$ the number on the secondary winding. Combining equations (3.23) gives the ratio of secondary voltage to primary voltage:

$$\frac{v_s}{v_p} = \frac{N_s}{N_p} = n$$

(3.24)

where $n$ is the turns ratio.

If $\Re$ is the reluctance of the core,

$$\Re\Phi = N_p i - N_s i$$

(3.25)

For an *ideal transformer* the reluctance of the core is zero, so that the ratio of the secondary current to primary current is given by

$$\frac{i_s}{i_p} = \frac{N_p}{N_s} = \frac{1}{n}$$

(3.26)

### 3.6.1 Impedance transformation

If a load $Z_L$ is connected across the secondary winding of an ideal transformer,

$$Z_L = \frac{v_s}{i_s} = \frac{n v_p}{i_p/n} = n^2 \frac{v_p}{i_p}$$

(3.27)

Hence the impedance seen at the primary side of the transformer is given by

$$Z_p = Z_L/n^2$$

(3.28)

### 3.6.2 A model for a 'real' transformer

The ideal transformer discussed above is only approximated by a real transformer. Some of the more significant effects that have been neglected are:

1. No losses have been included in the core or windings
2. The primary inductance has been assumed infinite (zero reluctance)
3. All flux passes through both the primary and secondary (no flux leakage)
4. No stray capacitance has been included.

A circuit model which takes into account effects (1)–(3) is shown in Figure 3.12. The losses in the windings due to finite resistance (copper losses) are represented by the resistances $R_p$ and $R_s$. At high frequencies these resistances may be greater than their d.c. values due to the skin effect, which forces the current to flow close to the surface of the wire, increasing its resistance. The finite reluctance of the core leads to an inductance $L_p$ shunting the input (the primary inductance). Only a fraction, $k$, of the flux in the primary is linked to the secondary, and this introduces series *leakage inductances*, $(1-k)L_p$ and $(1-k)L_s$, where $k$ is usually close to 1. The secondary and primary inductances, $L_s$ and $L_p$ , have a ratio proportional to the square of the turns ratio $n$. The core losses are represented by a resistance, $R_c$, across the shunt inductance. The value of $R_c$ will depend on frequency as it must take hysteresis and eddy current losses into account.

This model is not particularly easy to use and for a transformer with low flux leakage and small losses a simpler equivalent circuit may be sufficient. The secondary leakage inductance and series resistance may be moved to the primary side of the ideal transformer, provided the values are divided by $n^2$ (as required by the impedance scaling relation). If $k$ is close to 1, and $R_c$ is large, then the further simplification of moving $kL_p$ and $R_c$ to the input terminals is possible, as in Figure 3.13, where

$$L_m = L_p$$
$$L_e = L_p(1 - k) + L_s(1 - k)/n^2 \qquad (3.29)$$
$$R_e = R_p + R_s/n^2$$

This simplified circuit is often sufficient to estimate the equivalent series or shunt

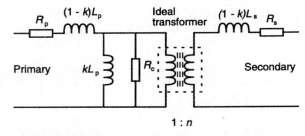

**Figure 3.12** *Circuit model for transformer with two windings*

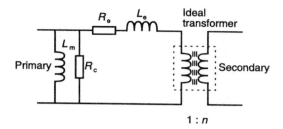

**Figure 3.13** *Simplified circuit model of transformer*

impedance imposed by a transformer. With this simplified model the primary current may be clearly separated into two components, the *magnetization current* flowing in $L_m$, and the load current, scaled by passing through the ideal transformer. The magnetizing current is independent of the load for this model.

A practical low-frequency power transformer is usually designed so that the maximum magnetic flux approaches the saturation value for the core when the primary voltage is at its rated value. This is desirable in terms of minimizing the size of the core required. A consequence is that the transformer should not be used with a primary voltage higher than its rated value, or the magnetization current will be increased and will drive the core into saturation. Lower voltages may be used but the power that may be transferred by the transformer will be reduced. The maximum flux in high-frequency transformers is usually limited by the need to limit the power dissipated in the core. The higher the frequency, the lower the maximum flux.

## 3.7 Ferrite cores

A variety of different designs of core are used depending on the application. A few styles used for ferrite cores are shown in Figure 3.14. There are generally two major considerations in choosing the shape: the ability of the core to contain the flux with minimal flux leakage and the need to provide adequate space for the windings. Toroidal cores probably have the lowest leakage if no air gap is required, but are rather more difficult to wind. Pot cores have low leakage even when an air gap is required, but are generally only available in smaller sizes. E–I or E–E cores are widely used for power transformers and energy storage inductors.

Manufacturers of ferrite cores, in addition to dimensioned drawings of core shapes, usually specify key dimensional constants for the core:

- The core factor or core constant $\Sigma l/A$, which determines the reluctance if the permeability is known
- The effective magnetic path length $l_e$
- The effective core cross-sectional area $A_e$
- The effective core volume, which specifies the energy storage, if there is no air-gap and the permeability and saturation flux are known.

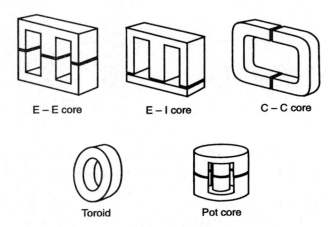

Figure 3.14  *Types of ferrite core*

The additonal factors that are required to calculate the magnetic properties of the core are the properties of the core material, and these are often specified in graphical form. A useful quantity often supplied is the *inductance factor*, $A_L$, sometimes called the specific inductance. This is the inductance of the core wound with a single turn, and is equal to the reciprocal of the reluctance (i.e. $1/\Re$). Hence, for a coil of $N$ turns on a core with a specific inductance $A_L$, the inductance is given by

$$L = A_L N^2$$

## Summary

This chapter has attempted to give an overview of some of the factors which affect the performance of inductors and transformers, and which must be taken into account in their use. The magnetic circuit model is the basis for understanding inductors using ferromagnetic cores. Many other practical issues have to be taken into account in the design of these components, for example the type shape and size of the core and the choice of wire (or even copper strip) for the windings. These issues are beyond what can be addressed here. The books listed in the References all give detailed information on the practical aspects of inductor and transformer design. Core manufacturers' data books give useful information on properties of materials, core characteristics and often basic design information.

## Self-assessment questions

3.1   A core made from two C-shaped sections has a total magnetic path length of 100 mm and a uniform cross-section 10 mm $\times$ 20 mm. Calculate the reluctance if the core has a relative permeability of 2500.

3.2 The core described in Question 3.1 is assembled with a non-magnetic spacer between the two halves of the core, so as to give two 0.5 mm air gaps. What is its new reluctance?

3.3 A coil of 25 turns is wound around the core described in Question 3.2. Calculate its inductance, and the maximum working current if the maximum flux density is 0.35$T$.

3.4 A pot core has an effective magnetic path length of 46 mm and an effective core area of 95 mm$^2$. If the relative permeability is 1800, what is the inductance factor, $A_L$, for the core?

3.5 The core described in Question 3.4 is modified by the inclusion of an air gap of width 0.2 mm in the centre pole of the core which has an area of 95 mm$^2$. The effective core volume is 4300 mm$^3$, and the maximum working flux is 0.35$T$. Estimate the maximum magnetic energy that may be stored in the core.

3.6 A small transformer is to be wound on a pot core with an inductance factor of 2000 nH. The primary inductance should be 10 mH, and the transformation ratio 2:1. How many turns are required for the two windings?

## Tutorial questions

**3.1** What are the key assumptions on which the magnetic circuit model is based?

**3.2** Explain why hysteresis losses in a magnetic core are proportional to frequency, while eddy current losses are proportional to the square of the frequency.

**3.3** Why must a steel core for a transformer be laminated while a ferrite core is not?

**3.4** Explain why inductors generally use a core with an air gap.

**3.5** In order to choose an appropriate core size for an energy storage inductor a key parameter is the volume of the air gap. Why?

**3.6** Why do transformers generally not need an air gap?

**3.7** What will limit the lowest frequency at which a transformer may be used?

**3.8** What will limit the maximum primary voltage that may be applied to a transformer?

**3.9** What are the principal sources of power dissipation in a transformer designed for use at 50 or 60 Hz?

## References

Lee, R., Wilson, L. and Carter, C. E., *Electronic Transformers and Circuits*, 3rd edition, John Wiley, New York, 1988.

McLyman, W. T., *Transformer and Inductor Design Handbook*, Marcel Dekker, New York, 1978.

McLyman, W. T., *Magnetic Core Selection for Transformers and Inductors*, Marcel Dekker, New York, 1982.

Mohan, N., Undeland, T. M. and Robbins, W. P., *Power Electronics; Converters, Applications and Design*, 2nd edition, John Wiley, 1995, Chapter 30.

CHAPTER 4

# *Rectification and smoothing*

## 4.1 Introduction

Rectification or the conversion of a.c. to d.c. is perhaps the longest-established, the simplest and the most familiar function performed by power electronics. The objective of this chapter is to analyze the performance of single-phase rectifiers and their associated smoothing filters. Two approximate analytic approaches will be considered which illustrate two methods of handling this type of circuit and the results will be compared with the results of circuit simulation.

Rectifiers are usually considered in two groups, half-wave rectifiers where current is drawn from the supply during one half of the cycle and full-wave rectifiers where both positive and negative half-cycles are utilized. The familiar basic single-phase half-wave rectifier circuit is shown in Figure 4.1. However, except at very low power levels, half-wave rectifiers are very rarely used. This is partly because they produce a larger ripple current in the load than do full-wave rectifiers, but, more seriously, the d.c. load current flows in the secondary of the transformer. This may be a transformer in the equipment or the distribution transformer of the supply company. Unless the transformer is designed to carry d.c., which is unusual, this may well cause saturation of the transformer core with associated power loss and distortion of the voltage waveform. Hence, half-wave rectifiers will be considered no further.

There are two basic full-wave rectifier circuits, shown in Figure 4.2: the bridge rectifier and the tapped transformer. The two circuits will produce the same voltage and current waveforms, but there are two significant differences. For the bridge rectifier circuit the load current always flows through two diodes effectively in series. The load voltage will therefore be less than the source voltage by the voltage lost across the two forward-biased diodes. This will reduce the efficiency

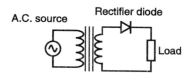

**Figure 4.1** *A single-phase, half-wave rectifier*

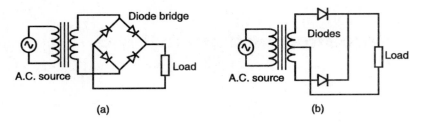

**Figure 4.2** *Single-phase full-wave rectifiers. (a) Bridge; (b) centre-tapped transformer*

significantly at low voltage when the diode forward drop is not much less than the load voltage. The tapped transformer has only one diode carrying current during either half-cycle, but the maximum reverse voltage across the diode not carrying current will have value of twice the peak load voltage (twice that for the bridge circuit). Thus the tapped transformer might be a better choice at a low-load voltage while the bridge circuit might be preferred at high voltage.

## 4.2  The full-wave rectifier with capacitance filter

A single-phase rectifier with a sinusoidal voltage source will produce a load voltage which is the modulus of a sinusoid. Usually what is required is a constant voltage. To achieve this, energy must be stored in a filter between the rectifier and the load. The simplest way in which this can be achieved is with a capacitance as shown in Figure 4.3. The capacitor stores energy while the rectifier voltage is near its peak and then supplies the load current for the rest of the cycle. For this reason it is usually called a *reservoir capacitor*. If the diode forward voltage is neglected then the load voltage will be as shown in Figure 4.4. The peak load voltage will be $\sqrt{2}V$, where $V$ is the r.m.s. value of the source voltage. During the conduction time, from $t_1$ to $t_2$, the capacitor will charge, while during the period $t_2$ to $t_3$ it will discharge. The greater the capacitance, the less will be the voltage ripple. However, increasing the capacitance will also reduce the conduction time of the diodes $(t_2-t_1)$, increasing the peak current.

The circuit may be analyzed using a piece-wise continuous approach. Assuming that the diode forward voltage may be neglected, and that the source has a low impedance, the circuit is represented by the two equivalent linear circuits, shown in Figure 4.5. Figure 4.5(a) shows the appropriate circuit while the diode is

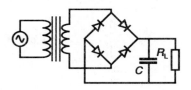

**Figure 4.3**  *A bridge rectifier with capacitance filter*

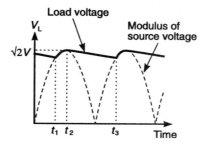

**Figure 4.4** *The voltage waveform of a full-wave rectifier with capacitance filter*

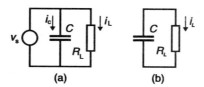

**Figure 4.5** *Equivalent circuits for a rectifier. (a) Diodes conducting; (b) diodes not conducting*

conducting and Figure 4.5(b) the appropriate circuit while the capacitor is supplying the load current.

If it is assumed that the load current is constant, then, while the diode is not conducting, the capacitor current is $-i_L$, and the charge lost from the capacitor is given by

$$q = i_L(t_3 - t_2) \tag{4.1}$$

and while the diode is conducting the current in the capacitor, $i_c$ is given by

$$i_c = C\frac{dv}{dt} = \sqrt{2}V\omega C \cos \omega t \tag{4.2}$$

where the source voltage has angular frequency $\omega$, and r.m.s. magnitude $V$. It is convenient to work in terms of the phase angle $\omega t$, rather than time, and to define an angle $\delta(t) = \omega t - \pi/2$. Substituting $\delta$ into equation (4.2) gives

$$i_c = -\sqrt{2}V \sin \delta(t) \tag{4.3}$$

Since conduction only occurs around the voltage maximum then equation (4.2) may be approximated to give

$$i_c \simeq -\sqrt{2}V\omega C\delta(t) \tag{4.4}$$

The capacitor current will therefore be as shown in Figure 4.6. While the diodes are not conducting the capacitor current will be constant at $-i_L$, and while the diode is conducting it will decrease at a constant rate with $i_c$ passing through zero at the

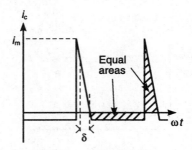

**Figure 4.6** *Current in the reservoir capacitor*

voltage maximum ($\omega t = \pi/2$). The maximum value of the capacitor current, $i_m$, will occur at $\omega t_1$, and the minimum value, $-i_L$, will occur at $\omega t_2$.

In order to find the peak current it is necessary to find the conduction angle. This is easily done since, in the steady state, the total charge supplied to the capacitor during the charging phase must equal the charge removed during discharge. Thus in Figure 4.6 the triangular area above the $i_c = 0$ axis while the capacitor charges must equal the trapezoidal area below the axis while the capacitor discharges. Making the simplifying assumption that the charging current terminates at the source voltage maximum, i.e. $\delta = 0$ (rather than at $\omega t_2$) the charge, $q$, stored during the charging phase is given by the area of the triangle, i.e.

$$q = i_m(\theta/2)T/(2\pi) \tag{4.5}$$

where $i_m$ is the peak charging current, $\theta$ is the conduction angle, i.e. ($\omega t_2 - \omega t_1$) and $T$ is the period ($2\pi/\omega$). The charge supplied by the capacitor to the load is found from the area below the $x$-axis,

$$q = i_L(\pi - \theta)T/(2\pi) \tag{4.6}$$

Combining equations (4.5) and (4.6) gives

$$\frac{i_m}{i_L} = \frac{2(\pi - \theta)}{\theta} \tag{4.7}$$

At time $t_1$, $\delta$ is equal to $-\theta$, so that from equation (4.4) a further equation linking $i_m$ and $\theta$ is obtained:

$$i_m = \sqrt{2}V\omega C\theta \tag{4.8}$$

Combining equations (4.7) and (4.8) gives an equation for $i_m$:

$$i_m^2 + 2i_m i_L - 2\pi\sqrt{2}V\omega C i_L = 0 \tag{4.9}$$

from which a value for $i_m$ is easily found:

$$i_m = i_L(\sqrt{1 + 2\pi\sqrt{2}V\omega C/i_L} - 1) \tag{4.10}$$

If $\theta$ is small compared with $\pi$, it follows that $i_m$ is much greater than $i_L$ (equation (4.7)), and hence

$$i_m = \sqrt{2\pi\sqrt{2}V\omega Ci_L} \qquad (4.11)$$

Substituting from equation (4.11) into equation (4.8) gives the conduction angle

$$\theta = \sqrt{\frac{\sqrt{2}\pi i_L}{V\omega C}} \qquad (4.12)$$

The current drawn from the supply while the rectifier is conducting is given by

$$i_s(\omega t) = i_c(\omega t) + i_L \qquad (4.13)$$

and is zero at other times. The r.m.s. value of the source current is given by

$$i_{rms} = \sqrt{\frac{1}{\pi}\int_0^\pi i_s^2(\omega t)\mathrm{d}(\omega t)}$$

$$\approx \sqrt{\frac{1}{\pi}\int_0^\theta \left(i_m\left(1 - \frac{\phi}{\theta}\right) + i_L\right)^2 \mathrm{d}\phi} \qquad (4.14)$$

$$= \sqrt{\frac{\theta(i_L^2 + i_L i_m + i_m^2/3)}{\pi}}$$

where $\phi = \omega t - \pi + \theta$ and $i_c$ is assumed to decreases linearly from $i_m$ to zero as $\phi$ varies from 0 to $\theta$.

The peak-to-peak ripple voltage is found from equation (4.6):

$$\Delta v = \frac{q}{C} = \frac{i_L(\pi - \theta)T}{2\pi C} \qquad (4.15)$$

### EXAMPLE 4.1

Consider a simple full-wave rectifier supplied from a 10 V r.m.s. sinusoidal a.c. supply. The filter capacitor has a value of 4.7 mF and the load resistance is 16 $\Omega$. Neglecting the source impedance and the diode forward voltage, estimate (a) the peak current from the source, (b) the r.m.s. source current and (c) the ripple voltage.

### SOLUTION

(a) Assuming that the load voltage is constant at the peak value, the load current $i_L$ will have a value of approximately $10\sqrt{2}/16$ A. From equation (4.11) the peak current may be estimated by

$$i_m = \sqrt{2\sqrt{2}\pi \times 10 \times 100\pi \times 4.7 \times 10^{-3} \times 0.88}$$
$$= 10.7 \text{ A}$$

(b)  The conduction angle deduced from equation (4.12) is 0.51 rad and the r.m.s. source current estimated from equation (4.14) is given by

$$i_{rms} = \sqrt{\frac{0.51 \times (0.88^2 + 0.88 \times 10.7 + 10.7^2/3)}{\pi}} = 2.8 \text{ A}$$

(c)  The ripple voltage is found from equation (4.15), and is given by

$$\Delta v = \frac{0.88 \times (\pi - 0.51) \times 20 \times 10^{-3}}{2\pi \times 4.7 \times 10^{-4}} = 1.6 \text{ V}$$

With the large peak current predicted by this simplified analysis it is obvious that the effect of even small values of source impedance will be significant. It would not be hard to include the impedance in the model, but an analytic solution, while it might be possible, would be much more complex. A simpler method of examining the effect of source impedance is to use circuit simulation.

## 4.3  SPICE simulation of a full-wave rectifier with capacitance filter

Circuit simulation has become a widespread and versatile tool used in the design and analysis of electronic circuits. Perhaps the most widely used circuit simulator is SPICE, which is now available in a variety of commercial forms. The version used in the examples to be included here was the Evaluation Version of PSPICE$^{TM}$ from the MicroSim Corporation, which runs on a PC. The simulations were run using a SPICE netlist to describe the circuit.

The rectifier and filter described in Example 4.1 has been simulated with a source resistance in series with the supply. The SPICE netlist is given in Appendix 1. Inspection of the model used for the diodes in this listing will show that the value used for the ideality factor, $N$, which would normally lie between 1 and 2, has been set to a very small value of 0.01. This has the effect of making the diode model behave very like an ideal diode with no forward voltage, so that the results may be compared with Example 4.1. The circuit in the listing differs from that in Example 4.1 in that a source resistance, RS, has been included.

The current flowing in source resistance is shown in Figure 4.7. A minimum source resistance of 5 m$\Omega$ was necessary to avoid instability in the integration, but with this very small value the current waveform is close to that expected with no source resistance, and the peak value is in reasonable agreement with that expected from the simple analysis as in Example 4.1. However, with a modest 0.2 $\Omega$ source resistance the peak current is significantly reduced.

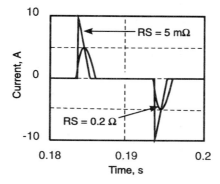

**Figure 4.7** *The source current with two values of source impedance*

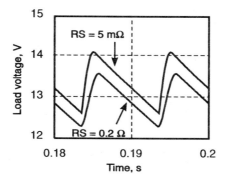

**Figure 4.8** *Ripple on the load voltage*

The load voltage is shown in Figure 4.8, indicating that the waveform and ripple with a very low source impedance is as expected. Increasing the source resistance reduces the mean load voltage without significantly changing the peak-to-peak ripple voltage.

It is clear that even a small source resistance will significantly change the peak current and the r.m.s. current drawn from the source. The values calculated ignoring the source impedance can only be viewed as an upper bound.

A second effect which is also frequently significant in practical power supplies is that where a transformer is used to supply the rectifier the secondary voltage waveform will be significantly distorted by the non-linearity of the transformer core. Small transformers such as are used in electronic equipment are operated very close to their saturation flux density and the voltage waveform frequently shows significant distortion in the form of flattening of the crests of the waveform. This has the effect of reducing the peak rectifier current and lengthening the conduction angle.

## 4.4 The full-wave rectifier with an inductance input filter

The rectifier with a capacitance filter leads to a poor current waveform on the source side of the rectifier and a large ripple current in the reservoir capacitance. If a low-pass filter with an inductance input is used, as in Figure 4.9, then the peak source current and the ripple current may be reduced substantially. With this type of filter there are two distinct modes of operation. If the load current is large then the current in the inductor will be continuous. This is the desirable mode of operation as it gives lower peak and r.m.s. source currents and better voltage regulation. At low-load current the current in the inductor becomes discontinuous. This separation into continuous and discontinuous modes of operation is common to rectifier and switching circuits employing inductors.

The circuit may be analyzed using the piece-wise continuous approach used for the capacitor filter. If, however, the current in the inductor is continuous then an alternative, and somewhat simpler, approach is possible. If the inductor current is continuous then the rectifier is always supplying current and the voltage across the rectifier must be the modulus of a source voltage, neglecting the diode forward voltage. This enables Fourier analysis to be used to find the magnitude of the various frequency components of the output of the rectifier to be calculated. Then, using linear circuit theory, the mean voltage and ripple voltage across the load and the ripple current in the capacitor may be calculated.

Neglecting the diode forward voltage, the voltage across the input to the filter is given by

$$v(t) = \sqrt{2}V|\sin \omega t| \tag{4.16}$$

Writing $v(t)$ as a Fourier series gives

$$v(t) = C_0 + \sum_{n=1}^{\infty}[C_n \cos n\omega t + S_n \sin n\omega t] \tag{4.17}$$

where

$$C_0 = \frac{1}{2\pi}\int_0^{2\pi} v(\omega t)\, \mathrm{d}(\omega t)$$

$$C_n = \frac{1}{\pi}\int_0^{2\pi} v(\omega t)\cos n\omega t\, \mathrm{d}(\omega t) \tag{4.18}$$

$$S_n = \frac{1}{\pi}\int_0^{2\pi} v(\omega t)\sin n\omega t\, \mathrm{d}(\omega t)$$

Considerable simplification is possible. The fundamental ripple frequency is $2\omega$, so that all the coefficients with odd $n$ are zero, and the function $v(t)$ is even, so all the coefficients $S_n$ are zero.

The mean value of the load voltage is given by

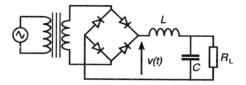

**Figure 4.9** *A bridge rectifier with an inductance input filter*

$$C_0 = \frac{2\sqrt{2}V}{\pi} = 0.9V \qquad (4.19)$$

and the fundamental ripple voltage is given by

$$C_2 = \frac{\sqrt{2}V}{\pi} \int_0^{2\pi} |\sin \omega t| \cos 2\omega t \, d(\omega t)$$

$$ \qquad (4.20)$$

$$= -\frac{4\sqrt{2}V}{3\pi}$$

Higher-order harmonics will be neglected, both because they are smaller and because they are attenuated by the filter.

If it is assumed that $C$ is sufficiently large that the ripple voltage across the load is small compared with the ripple voltage at the input of the filter, then the amplitude of the fundamental component of the ripple current $i_r$ in the filter inductor is given by

$$i_r = \frac{C_2}{2\omega L} = \frac{2\sqrt{2}V}{3\pi\omega L} \qquad (4.21)$$

If the current is to be continuous then the mean load current must exceed the amplitude of the ripple current, otherwise the instantaneous current will fall to zero and become discontinuous (it cannot become negative). If the load resistance is $R_L$ and the condition for continuous current in the inductor is $i_r < i_L$, then the condition for continuous conduction may be written

$$\frac{2\sqrt{2}V}{3\pi\omega L} < \frac{2\sqrt{2}V}{\pi R_L}$$

or

$$\frac{R_L}{\omega L} < 3 \qquad (4.22)$$

If this condition is satisfied then the current in the conductor will be continuous and the linear analysis is valid. In the limit of very large inductance the current in the inductor becomes constant, in which case the source current will become a square-wave, with an amplitude equal to the load current, $I_L$. Hence the r.m.s. source current will also be $I_L$.

**EXAMPLE 4.2**

A full-wave rectifier has an inductance input filter, comprising a 50 mH inductance and a 4.7 mF capacitance. The power supply is sinusoidal with a frequency of 50 Hz and an r.m.s. magnitude of 25 V. What is the critical load current for the inductor current to be continuous? If the load current exceeds the critical current, estimate the output ripple voltage.

**SOLUTION**

The amplitude of the ripple current in the inductor is given by equation (4.21):

$$i_r = \frac{2\sqrt{2}V}{3\pi\omega L} = \frac{2\sqrt{2}.25}{3.10\pi^2.50.10^{-3}} = 0.478 \text{ A}$$

Provided the load current exceeds this value, the current in the inductor will be continuous. Hence the critical current for the inductor current to be continuous is 0.478 A.

The ripple current will split between the load and the capacitor. The capacitor has a reactance of only 0.34 $\Omega$ at 100 Hz, hence it is reasonable to assume that all the ripple current flows in the capacitor, hence the amplitude of the ripple voltage $v_r$ is given by

$$v_r = \frac{i_r}{2\omega C} = 0.162 \text{ V}$$

The amplitude of the ripple voltage will therefore be 0.16 V, or the peak-to-peak value 0.32 V, provided the load current exceeds the critical value.

## 4.5  The transient response of the inductance input filter

If the current in the inductor is continuous then the inductance and capacitance act as a series resonant circuit, damped by the effective series resistance of the inductor and capacitor, by the source resistance, and the resistance of the load across the capacitor. Changes in input voltage or load current may excite ringing in the filter at an angular frequency of $1/\sqrt{LC}$, unless the damping of the resonance is large.

The circuit described in Example 4.2 has been simulated to show the initial start-up transient. The mean load voltage from equation (4.19) is 22.5 V, so that a load resistance of 22 $\Omega$ gives a mean load current of about 1A. This is sufficient to ensure that, in the steady state, the inductor current is continuous. The SPICE listing is given in Appendix 1.

Figure 4.10 shows the load voltage (a) and the load current (b) when the circuit is first turned on. The start-up transient shows a very large overshoot, with the peak voltage reaching almost double the mean load voltage. At this point the current in the inductor falls to zero, and the load voltage then decays as the capacitor

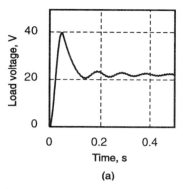
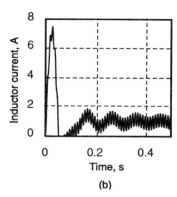

**Figure 4.10** *A rectifier with an inductance input filter and a continuous inductor current. (a) Load voltage; (b) inductor current*

discharges. When the load voltage falls far enough for the inductor current to flow again the start-up transient decays away as a damped oscillation about the mean load voltage.

## 4.6  Discontinuous current with an inductance input filter

At low-load current the inductor current will become discontinuous and the load voltage will rise above the average value of the rectified sine-wave, $0.9V$. With no load the voltage will rise to the peak voltage from the rectifier of $\sqrt{2}V$. The Fourier method described above cannot be used to analyze the case of discontinuous conduction, because the voltage waveform cannot be determined. The input to the filter follows the modulus of the source voltage only while the diodes are conducting. When the diodes stop conducting the voltage at the input of the filter will jump to the voltage across the capacitor. The time at which this jump occurs can only be found if a solution for the inductor current has already been obtained. The piece-wise integration method can be used, but an analytic solution is not straightforward. The more convenient method of analysis is obviously circuit simulation.

Using the same netlist as in Section 4.5, but with the load resistance increased to $100\ \Omega$, the circuit was simulated for 0.5 s. The same start-up transient was observed, with a slightly larger maximum voltage as shown in Figure 4.11. The decay of the transient is slower because of the larger time constant with the larger load resistance. This initial transient can only be avoided if the resonant frequency of the filter is well above the ripple frequency, which, of course, defeats the object of using the filter. When the transient has settled, the voltage waveform at the input of the filter is as shown in Figure 4.12. A point to notice is the overshoot when the voltage across the inductor jumps up to the load value. In this simulation this is an artifact of the simulation, but in a real circuit, or with more careful modelling, a

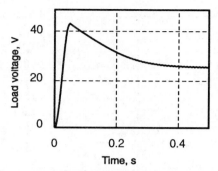

**Figure 4.11** *A rectifier with an inductance input filter and a discontinuous inductor current, load voltage*

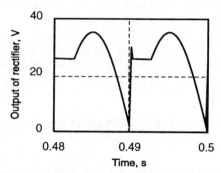

**Figure 4.12** *A rectifier with an inductance input filter and a discontinuous current, voltage at input of the filter*

similar effect would be observed due to the stray capacitance of the inductor and the capacitance of the diode (not included in this model) which will resonate with the inductance, leading to an oscillatory transient as the conduction ceases.

## 4.7 Power factor

For a sinusoidal voltage source driving a linear load the current will also be sinusoidal, but with a phase shift $\phi$. The power delivered to the load, averaged over one cycle, will be given by

$$P = VI \cos \phi \qquad (4.23)$$

where $V$ is the r.m.s. voltage and $I$ the r.m.s. current. The quantity $\cos\phi$ is frequently defined to be the 'power factor', since it relates the true power delivered to the load, to the 'VA' product, i.e. the product of the r.m.s. voltage and the r.m.s. current. Thus the power factor, PF, could be defined as

$$PF = \frac{\text{Power delivered}}{\text{VA product}}$$

For a sinusoidal source and linear load this would agree with the more familiar definition. However, this second definition may be applied when the load is non-linear and the load current is consequently non-sinusoidal, or indeed the source voltage may also be non-sinusoidal. If the load is resistive then the true power and the VA product will always be equal and the power factor will be unity.

If the source voltage, $v(t)$, is a simple sinusoid of magnitude $V$, and the load is non-linear, then the load current, $i(t)$, may be written as a Fourier series:

$$v(t) = \sqrt{2}V \sin \omega t$$

$$i(t) = \sqrt{2} \sum_{n=1}^{\infty} (I_{cn} \cos n\omega t + I_{sn} \sin n\omega t)$$

where $I_{cn}$ and $I_{sn}$ are the r.m.s. magnitudes of the in-phase and quadrature Fourier components at frequency $n\omega$ (the r.m.s. magnitude is equal to the amplitude/$\sqrt{2}$), and it has been assumed that there is no d.c. component. Multiplying $i(t)$ by $v(t)$ gives the instantaneous power, which may then be integrated over one cycle to give the average power:

$$P = \frac{1}{2\pi} \int_0^{2\pi} v(\omega t) i(\omega t) \mathrm{d}(\omega t)$$

$$= \frac{V}{\pi} \int_0^{2\pi} \sin (\omega t) \left( \sum_{n=1}^{\infty} (I_{cn} \cos \omega t + I_{sn} \sin \omega t) \right) \mathrm{d}(\omega t) \qquad (4.24)$$

$$= V I_{s1}$$

where all the integrals are zero, except the term involving $\sin^2(\omega t)$. The term $I_{S1}$ is the r.m.s. value of the fundamental Fourier component of the current which is in phase with the source voltage.

The power depends only on the in-phase component of the fundamental of the load current. The higher-frequency components (harmonics), generated in the current by the non-linearity of the load, do not contribute to the power. However, they do contribute to the r.m.s. value. Calculating the r.m.s. value of the current gives

$$I = \sqrt{\frac{1}{2\pi} \int_0^{2\pi} i^2(\omega t) \mathrm{d}(\omega t)}$$

$$\qquad (4.25)$$

$$= \sqrt{\sum_{n=1}^{\infty} (I_{cn}^2 + I_{sn}^2)}$$

where the current has been expressed as a Fourier series before the integration is performed. Thus the power factor is given by

$$PF = \frac{I_{s1}}{\sqrt{\sum_{n=1}^{\infty}(I_{cn}^2 + I_{sn}^2)}} \qquad (4.26)$$

An alternative way of writing equation (4.26) is to express the power factor in terms of the r.m.s. value of the fundamental component of the current, i.e.

$$PF = \frac{I_{1rms}}{I_{rms}} \cos \phi_1 \qquad (4.27)$$

where $I_{1rms}$ is the r.m.s. value of the fundamental component of the source current, $I_{rms}$ is the r.m.s. value of the source current and $\phi_1$ is the displacement angle of the fundamental component of the current. A non-linear load which generates harmonics will have a poor power factor, as will a reactive load where the quadrature part of the fundamental component of the current is significant.

### EXAMPLE 4.3
A full-wave rectifier with a simple capacitor filter is supplied from a low source impedance with a sinusoidal voltage of magnitude 20 V r.m.s. The reservoir capacitance has a value of 20 mF and the load is 10 $\Omega$. Estimate the power factor of the load presented by the rectifier to the source, neglecting the power loss in the diodes.

### SOLUTION
Assuming that the forward voltage of the diodes may be neglected, and neglecting the ripple voltage, the load voltage will be $20.\sqrt{2} = 28.3$ V, and the load current 2.8 A. Using equation (4.12) the conduction angle $\theta$ may be estimated:

$$\theta = \sqrt{\frac{\sqrt{2}\pi i_L}{V\omega C}} = \sqrt{\frac{\sqrt{2}\pi \times 2.8}{20 \times 100\pi \times 0.02}} = 0.31 \text{ rad}$$

The maximum current in the capacitor, $i_m$, may be estimated from equation (4.11):

$$i_m = \sqrt{2\pi\sqrt{2}V\omega Ci_L} = \sqrt{2\pi\sqrt{2} \times 20 \times 100\pi \times 0.02 \times 2.8} = 56 \text{ A}$$

From $i_m$ and $\theta$, the r.m.s. value of the source current may be calculated from equation (4.14) and is given by

$$I = \sqrt{\frac{\theta(i_L^2 + i_L i_m + i_m^2/3)}{\pi}}$$

$$= \sqrt{\frac{0.31 \times (2.8^2 + 2.8 \times 56 + 56^2/3)}{\pi}}$$

$$= 10.9 \text{ A}$$

The power from the source must equal the power delivered to the load, since losses have been neglected. The power from the source is therefore $28.3^2/10 = 80$ W. Hence the power factor is given by

$$\text{PF} = \frac{80}{20 \times 10.9} = 0.37$$

This will be a pessimistic estimate of the power factor, since the peak current and the r.m.s. current will be reduced by the effect of the source impedance.

In Section 4.3 SPICE was used to calculate the waveforms of the input voltage for a rectifier with source resistances of 0.005 Ω and 0.2 Ω. SPICE also allows for the Fourier frequency components to be calculated from the waveform. For the example discussed in Section 4.3 the frequency spectra of the source currents are as shown in Figure 4.13. The harmonic components have been expressed as the r.m.s. magnitudes, as used in equations (4.24)–(4.26), not the conventional amplitudes. It can be seen that both spectra are rich in harmonics, especially that for the rectifier with the lower source impedance. Notice that the symmetry of the waveform means that the even harmonics do not appear. The r.m.s. line current and the power factor may be calculated from these spectra using equations (4.25) and (4.26). For the low source resistance the r.m.s. line current is about 2.3 A and the power factor 0.5. The

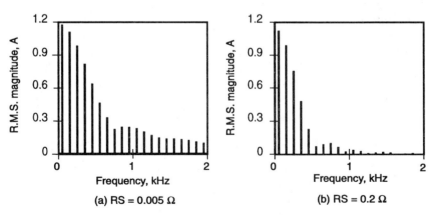

**Figure 4.13** *The harmonic content of a rectifier with capacitive filter with (a) 0.005 Ω source resistance and (b) 0.2 Ω source resistance*

r.m.s. current is somewhat less than that calculated in Example 4.1 (2.8 A), probably due to the approximations made in the analytic approach. For the 0.2 $\Omega$ source resistance, the r.m.s. line current is calculated to be 1.77 A and the power factor 0.63.

For a rectifier with a capacitance input filter the power factor is always poor because the discontinuous pulses of current lead to a high r.m.s value. The inductor input filter is somewhat better. If the inductor is very large then the input current is a square wave of amplitude $I_L$, where $I_L$ is the load current. It follows that the r.m.s. input current is also $I_L$, so the VA product is $VI_L$ where $V$ is the r.m.s. source voltage. The power delivered to the load, neglecting the diode losses, is just $0.90VI_L$, which must equal the input power, hence the power factor is 0.90.

## 4.8 The voltage doubler

Voltage-multiplying rectifiers are used to generate a d.c. voltage higher in value than the peak of the a.c. supply voltage. Multiplier circuits such as those shown in Figure 4.14 have been used for many years to generate high-voltage low-current d.c. supplies. The circuit of Figure 4.14(a) is a voltage doubler. At low-load current the output is twice the peak input voltage, while the circuit of Figure 4.14(b) is a voltage tripler whose output voltage is three times the peak input voltage, at low load current. These circuits may be extended to obtain higher multiples, but, apart from the voltage doubler, are generally only used where the load current is small, e.g. to generate the high voltage for a CRT in an oscilloscope.

To understand the operation of these circuits, consider the voltage-doubler circuit of Figure 4.14(a). If the load current is low, the capacitor $C1$ will charge so that it has a d.c. potential across it equal to the peak supply voltage. When the supply voltage is at its maximum negative value the potential at node 2 will be zero, $C1$ will charge through $D1$ during this half-cycle. When the source voltage is close to its maximum positive value node 2 will be at a potential of twice the peak supply voltage and $C2$ will charge to this potential through $D2$. Note that since power is drawn from both half-cycles this doubler is essentially a full-wave rectifier.

Power supplies for use in electronic equipment and computers now frequently avoid the use of a power-frequency transformer. The approach is to rectify the power directly from the wall socket and then use a high-frequency isolating

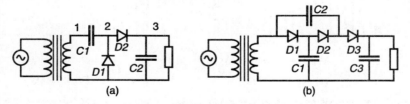

**Figure 4.14** *Voltage multipliers. (a) Voltage doubler; (b) voltage tripler*

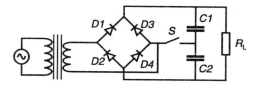

**Figure 4.15** *A 'Universal rectifier', a full-wave bridge rectifier or voltage doubler*

converter to change the voltage to the required value and provide regulation. Such power supplies are referred to as off-line switch-mode power supplies (SMPs). Such supplies are frequently required to work from line voltages in the range 100–120 V and also 200–240 V, depending on the country in which it is to operate. With equipment using a power transformer at the input a simple switch, or link, may be used to select the operating voltage. With SMPs either the switching regulator must cope with a very wide input voltage range or another solution must be found. The alternative solution is shown in Figure 4.15. If the switch $S$ is open then the rectifier operates as a bridge rectifier in the usual way. If the switch $S$ is closed then the circuit operates as a voltage doubler. In this second mode of operation the diodes $D3$ and $D4$ take no active part. The diodes $D1$ and $D2$ act as half-wave rectifiers charging $C1$ and $C2$ on opposite half-cycles, charging each capacitor to the peak line voltage, and giving an output voltage of twice the peak line voltage. Thus the rectifier may be used as a bridge with a 200–240 V line and as a doubler with a 100–120 V line.

## Summary

This chapter has examined the properties of simple single-phase rectifier circuits. Two methods of analyzing such circuits using essentially linear methods have been illustrated. The piece-wise linear approach can be used with a variety of circuits when the non-linearities are essentially local discontinuities, joined by regions of linear operation. The technique will be used to analyze the operation of switching power supplies and controlled rectifiers. The second approach using Fourier analysis is of more restricted application and can only be used when the voltage (or current) waveform is known, as it is with a rectifier driving an inductive load with a continuous current. This technique is also useful with controlled rectifiers.

It has been shown that rectifiers frequently provide a load which has a poor power factor and the current drawn from the source is rich in harmonics.

## Self-assessment questions

4.1   A single-phase bridge rectifier is supplied from an a.c. source with a frequency of 50 Hz and a magnitude of 100 V r.m.s. It supplies a mean load current of 10 A to a resistive load. What will be the minimum values for the diode ratings (a) for the reverse voltage and (b) for the average forward current (averaged over one cycle)?

4.2    A single-phase bridge rectifier with a simple capacitance filter has a sinusoidal input voltage of 220 V r.m.s. at 50 Hz and supplies an average load current of 0.1 A. The reservoir capacitor has a value of 100 $\mu$F. Estimate the diode conduction angle, neglecting the source resistance and diode forward voltage. What will be the amplitude of the peak-to-peak ripple voltage?

4.3    For the rectifier described in Question 4.2, estimate the peak input current and the r.m.s. source current.

4.4    A bridge rectifier with a source voltage of 100 V r.m.s. at a frequency of 60 Hz and a reservoir capacitor of 1 mF supplies a load with a resistance of 250 $\Omega$. Calculate the average and r.m.s. values of the current in each diode of the bridge rectifier, assuming that the source resistance may be neglected.

4.5    A single-phase bridge rectifier has an inductive input filter with an inductance of 0.1 H and a capacitance of 5000 $\mu$F. The source voltage has an r.m.s. magnitude of 220 V and a frequency of 60 Hz. What is the critical load resistance and corresponding critical load current at which the inductor current becomes continuous? Estimate the peak-to-peak ripple voltage across the load while the inductor current is continuous.

4.6    A single-phase bridge rectifier with a sinusoidal source of angular frequency $\omega$ and r.m.s. magnitude $V$ supplies current to a filter with an input inductance, $L$, and a large capacitor which keeps the load voltage constant. Assuming that the current in the inductor is continuous, calculate the amplitude of the ripple current at a frequency of $n\omega$, where $n$ is an integer, greater than 1.

4.7    A full-wave single-phase rectifier with a source voltage of 50 V r.m.s. at a frequency of 50 Hz has a reservoir capacitor of 10 mF and a load current of 2 A. Assuming that the source resistance and diode forward voltage may be neglected, estimate the VA product and the power factor of the power drawn from the source.

4.8    A voltage doubler using the circuit shown in Figure 4.15, with switch $S$ closed, operates from a supply of 110 V r.m.s. at 60 Hz. The capacitors $C1$ and $C2$ each have a value of 200 $\mu$F, and the load current is 0.1 A. Estimate the peak-to-peak ripple voltage across each capacitor, assuming that the conduction angle is very small. Sketch the voltage across $C1$ and the voltage across $C2$. Hence (or otherwise) estimate the peak-to-peak ripple voltage across the load.

4.9    A power supply is required to provide a voltage of 100 V d.c. at 0.1 A. A simple capacitive filter is to be used, and the ripple voltage must not exceed 5 V. The source is to be a single-phase 50 Hz supply. Neglecting the source impedance, calculate a suitable supply voltage (r.m.s.) and a value for the reservoir capacitor. Estimate the r.m.s. current in the capacitor. Explain qualitatively how the load voltage, r.m.s. current and ripple voltage will change if the source has a finite impedance.

# Tutorial questions

**4.1** Sketch the load voltage for a full-wave rectifier with a capacitance input filter.

**4.2** Sketch the source current and capacitor current for a full-wave rectifier with a capacitance input filter.

**4.3** Explain why the tapped transformer version of the full-wave rectifier may be preferred to the bridge rectifier at low-load voltage, while at high voltage the reverse is true.

**4.4** Explain why for a full-wave rectifier with a highly inductive load the input current will be a square wave.

**4.5** Why will an inductance input filter generally give a poor transient response to changes in load current?

CHAPTER 5

# *D.C. linear regulators*

## 5.1 Introduction

The simple rectifier circuits considered in Chapter 4 are examples of unregulated supplies. The load voltage will change in response to changes in the voltage of the source or the current in the load. To stabilize the load voltage such supplies are frequently used in conjunction with a voltage regulator. This will keep the load voltage constant as the a.c. source voltage or the load current is varied and will also reduce the ripple voltage across the load. Usually regulators are designed to provide a constant load voltage, since that is what is required for most electronic equipment. Regulators may also be used to stabilize the current where this is required by the load. An example of a load which requires current stabilization would be an arc or gas discharge device where the load voltage is almost independent of current, and the current must be held constant to control the power.

In this chapter we shall consider only linear regulators and only the regulation of direct current or voltage. The term 'linear regulator' is used to describe an electronic regulator in which the voltage or current is controlled using transistors and other active devices as variable impedance elements. This is in contrast to switching regulators (to be considered in the next few chapters), which use the transistors as switches that are either 'on' or 'off', with the current or voltage controlled by pulse-width modulation.

## 5.2 Shunt and series regulators

Regulators can be divided into two groups: shunt and series. A shunt regulator has a variable impedance element in parallel with the load (Figure 5.1(a)). The element diverts current from the load, and assuming that the source has a suitable impedance, the load voltage may be controlled by varying the amount of current diverted. When used to regulate voltage, a shunt regulator is a two-terminal device, since only two connections are needed to sense the load voltage and to sink the excess current. The series regulator has a variable-impedance element connected in series between power source and load (Figure 5.1(b)). The voltage across the load is sensed and the impedance of the series element varied so as to keep the load voltage constant. When used as a voltage regulator the series regulator is a three-

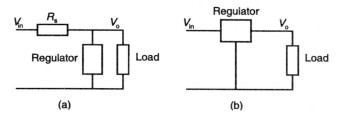

**Figure 5.1** *Regulator configurations. (a) Shunt; (b) series*

terminal device or circuit, since it must have a series element with two terminals and a third connection to sense the voltage across the load, as in Figure 5.1(b). It is worth noting that when used as a current regulator the shunt regulator becomes a three-terminal device while the series regulator requires only two terminals.

If the power source is basically a voltage source, such as a rectifier, then the series regulator is the preferred configuration since it is more efficient, especially at low-load current. Shunt regulators are generally used only in low-power circuits such as voltage references, (for example, voltage reference diodes or bandgap reference integrated circuits). An example where shunt regulators may be preferred is in controlling the voltage from solar cells, which behave as a current limited source, and greater overall efficiency is possible with a shunt regulator.

## 5.3 Definition of stability parameters

The performance of a voltage regulator is determined by its ability to maintain a constant load voltage as both load current and source voltage vary. Several figures of merit are defined to enable the performance to be quantified.

The *line regulation* is defined by

$$\text{line regulation} = \frac{\Delta V_0}{V_0} \times 100\%$$

where $\Delta V_o$ is the maximum change in the output voltage $V_o$ as the source (line) voltage varies over its specified range, with a constant load current. The *stabilization factor* is given by

$$\text{stabilization factor} = \left(\frac{V_{\text{in}}}{V_0}\right) \frac{dV_0}{dV_{\text{in}}}$$

where $V_{\text{in}}$ is the input voltage. Both these figures of merit express a way of measuring the ability of the regulator to reduce the effects of variations in the input voltage. The line regulation gives the maximum possible variation in output voltage, provided the input range specified is not exceeded, while the stabilization factor gives the ratio of the fractional change in the output voltage to the fractional change in the input. These are very different measures, and which is more appropriate depends on circumstances.

For the ability to keep the load voltage constant with changes in load current, the figures of merit are the *load regulation* and the *output resistance*, given by

$$\text{load regulation} = \frac{\Delta V_0}{V_0} \times 100\%$$

$$\text{output resistance} = -\frac{dV_0}{dI_0}$$

where $\Delta V_0$ is the maximum change in the output voltage as the load current varies over its full range, with constant line voltage. The expressions for the line and load regulation look the same. The difference is that in the expression for line regulation the change in load voltage is due to a change in the line voltage, while for load regulation the change in load voltage is due to a change in the load current.

## 5.4  The shunt regulator

The simplest type of voltage regulator uses the reverse breakdown of a silicon junction diode. Diodes made for this purpose are usually referred to as Zener diodes, although this nomenclature is rather misleading since the voltage is controlled by Zener breakdown only in devices rated at less than about 5 V. At higher voltage ratings the reverse breakdown voltage is determined by avalanche breakdown. The breakdown voltage due to the Zener mechanism has a negative temperature coefficient, while breakdown voltage due to avalanche multiplication has a positive coefficient. At a breakdown voltage of about 5 V both effects are active and the net temperature coefficient is close to zero. The characteristics of a Zener diode are shown in Figure 5.2. When the reverse voltage exceeds a critical value the reverse current increases rapidly with a *slope resistance* $\rho$. In the breakdown region the characteristic may be modelled as a voltage source $V_r$ in series with a resistance $\rho$.

The circuit of a simple shunt regulator using a Zener diode is shown in Figure 5.3, together with its small-signal equivalent. The load voltage, $V_o$, is given by

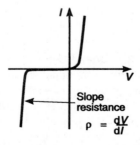

**Figure 5.2** *Characteristics of a Zener diode*

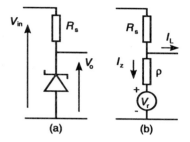

**Figure 5.3** *Shunt regulator with a Zener diode. (a) Circuit; (b) model*

$$V_0 = V_r + I_z\rho \tag{5.1}$$

where $I_z$ is the current in the Zener diode. The input voltage is given by

$$V_{in} = V_0 + (I_z + I_L)R_s \tag{5.2}$$

where $I_L$ is the load current. Using equations (5.1) and (5.2) the load voltage may be expressed as a function of the input voltage $V_{in}$ and the load current:

$$V_0 = \frac{V_r + V_{in}\rho/R_s - I_L\rho}{1 + \rho/R_s} \tag{5.3}$$

The output resistance, $R_o$, of the regulator is easily found by differentiating $V_0$ with respect to $I_L$ to give

$$R_0 = -\frac{\partial V_0}{\partial I_L} = \frac{\rho R_s}{\rho + R_s} \tag{5.4}$$

and the stabilization factor, $S$, found by differentiating $V_0$ with respect to $V_{in}$ is given by

$$S = \frac{V_{in}}{V_0}\frac{\partial V_0}{\partial V_{in}} = \frac{V_{in}}{V_0}\frac{\rho}{\rho + R_s} \tag{5.5}$$

In order to obtain a good stabilization factor $R_s$ must be much greater than the slope resistance $\rho$, in which case the output resistance of the regulator will be equal to $\rho$. However, the value of the series resistance must be such that the relation

$$I_L \le \frac{V_{in} - V_0}{R_s}$$

is satisfied otherwise the shunt regulator current will fall to zero. The maximum load current, $I_m$, is the value for which the equality condition is satisfied, at the minimum source voltage $V_{min}$. Hence $I_m$ is given by

$$I_m = \frac{V_{min} - V_0}{R_s} \tag{5.6}$$

This equation may be used to find the value of $R_s$ provided $V_{min}$ is known. For reasonable stability $V_{min}$ must be substantially greater than $V_0$.

Since the load voltage is almost constant, as the load current varies, the Zener current must also vary in such a way as to keep the total source current constant. The power supplied by the source is therefore constant at constant source voltage, irrespective of the load current. If the efficiency of the regulator ($\eta$) is defined as the ratio of power delivered to the load to the power supplied by the source then $\eta$ may be written

$$\eta = \frac{V_0 I_L}{V_{in} I_S} = \frac{V_0 I_L R_S}{V_{in}(V_{in} - V_0)} \tag{5.7}$$

The efficiency is clearly greatest when the load current has its maximum value.

### EXAMPLE 5.1

A series regulator uses a Zener diode with a nominal voltage of 4.7 V and a slope resistance of 1 $\Omega$. The source voltage is 10 V, and the load current may vary between 0 and 200 mA. Choose a suitable value for the series resistance and calculate the maximum power dissipated in the resistance and the Zener diode. Estimate the stabilization factor and the output resistance.

### SOLUTION

The current in the series resistance must exceed the maximum load current of 0.2 A and drops a nominal voltage of 5.3 V, which will vary somewhat as the load current varies. This gives a maximum resistance of 26.5 $\Omega$. Since the current in the resistance must exceed the value of 0.2 A, choose a value of 22 $\Omega$.

With this value the total current drawn from the source will be 5.3/22 = 0.24 A. Hence the power dissipated in the resistance will be $22 \times 0.24^2 =$ 1.3 W. The maximum power dissipated in the Zener diode occurs when the load current is zero and all the source current flows in the diode. This gives a power dissipation of $4.7 \times 0.24 = 1.13$ W.

The stabilization factor is given by equation (5.5) and substituting for $V_{in}$, $V_0$, $\rho$ and $R_s$ gives a stabilization factor of $(10/4.7) \times 1/(1+22) = 0.093$. Similarly, the output resistance is given by equation (5.4) as $\rho$ and $R_s$ in parallel, i.e. 0.96 $\Omega$

## 5.5  The series regulator

Series regulators have a variable impedance element between the source and load. As the load current or source voltage varies, the impedance of the series element (usually a transistor) is adjusted to keep the load voltage constant. The only current in the series element is the load current, so that neglecting the power required for the circuit to control the series element the efficiency is given by

$$\eta = \frac{I_L V_0}{I_L V_{in}} = \frac{V_0}{V_{in}} \tag{5.8}$$

In contrast to the shunt regulator, the efficiency does not depend on the load current, only on the ratio of load-to-source voltage. It is for this reason that series regulators are generally preferred unless the power is very low.

The series regulator is often based upon the simplified circuit of Figure 5.4. The transistor $Q1$ is the variable series element, usually referred to as the pass transistor. It has a base current supplied by the current source $I1$. The transconductance amplifier compares the output voltage, $V_0$, with a reference voltage, $V_R$, and diverts current from the base of $Q1$ so as to control the load voltage. Provided the gain is large, the load voltage is given by

$$V_0 \approx \frac{V_R(R_1 + R_2)}{R_2}$$

The transistor $Q2$ limits the load current. When the voltage across $R$ is of the order 0.6 V $Q2$ turns on, diverting current from the base of $Q1$ and limiting the load current. Current limiting is used to protect both the power supply and the load.

The simple current limiting circuit of Figure 5.4 has a major disadvantage in that if the load is short circuit, the pass transistor has to carry the limiting current while its collector voltage is the full source potential. If the supply is to sustain this condition the pass transistor and its heatsink must be able to withstand this power. In integrated circuit regulators it is common to have a circuit which will shut down the current if the temperature of the device rises outside permitted limits, providing an extra protective measure against sustained short-circuit load conditions. An alternative approach is to use *foldback current limiting*, which progressively reduces the limiting current as the load voltage falls. A simple circuit to implement this is shown in Figure 5.5.

The operation of this circuit is similar to the previous circuit, although note that the pass transistor is now a *p–n–p* type. In order to understand how the current

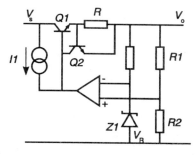

**Figure 5.4** *Simplified circuit of series regulator with current limiting*

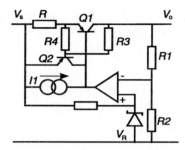

**Figure 5.5** *Simplified series regulator with foldback current limiting*

limiting circuit operates, consider the voltage across the base emitter junction of
$Q2$, which is given by

$$V_{\text{be2}} = -I_S R + \frac{R_4(V_0 - V_S)}{R_3 + R_4} \tag{5.9}$$

If the transistor turns on sufficiently to limit the current when the base–emitter
voltage of $Q2$ is $-V_t$ , then the condition for the limiting current is obtained from
Equation 5.9 and is given by

$$I_S \approx \frac{1}{R}\left(V_t + \frac{V_0 R_4}{R_3 + R_4} - \frac{V_S R_4}{R_3 + R_4}\right) \tag{5.10}$$

This is a simplistic view, which assumes that the transistor turns on at a well-
defined voltage, but it can be seen from equation (5.10) that as the load voltage
falls, the limiting current also falls, limiting the power dissipation in the pass
transistor.

Series regulators with over-current and thermal protection are available as
integrated circuits in a single power package. They are available as fixed voltage
and variable regulators in current ratings from 0.1 A up to about 5 A, and for either
positive or negative supplies. Where higher currents are needed, ICs are available
to provide the control and protection functions with external pass transistors.

## 5.6 Simulation of a series regulator with foldback limiting

In order to clarify the operation of a series regulator with foldback limiting,
consider the circuit shown in Figure 5.6. This is the complete circuit of a relatively
simple regulator with foldback limiting. The pass transistor Q1 is driven by a
common collector stage, or emitter follower, Q2. This combination, often called a
'Darlington pair', ensures adequate current gain to control the load current. The
feedback amplifier, which is formed by the long-tailed pair Q4 and Q5, compares a
reference voltage derived from a Zener diode with a fraction of the load voltage.
The transistor Q3 limits the load current and provides foldback limiting. This

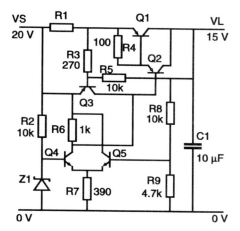

**Figure 5.6** *Circuit of regulator used for simulation*

circuit has been simulated using the SPICE netlist given in Appendix 1 (Section A1.3).

To demonstrate how the load voltage varies with load current the load comprises a voltage source VL in series with a load resistance RL of 1 Ω. The voltage source is swept from 15 V down to −2 V in 1 s. This gives a voltage current characteristic as shown in Figure 5.7. This clearly shows the low-output impedance until the current limits, then the characteristic folds back on itself as the voltage falls. The output impedance of the power supply is about 100 mΩ before the current limits.

If an a.c. current source is connected across the output then the output impedance can be obtained as a function of frequency. Figure 5.8 shows the output impedance for three values of the load capacitance C1. The output impedance at high frequency clearly depends upon the output capacitance C1. The regulator is an example of a feedback system, and peaks at about 100 kHz and 300 kHz arise from phase shift in the feedback amplifier, which could lead to instability, although such behaviour was not found when simulating this circuit. Provided C1 is sufficiently

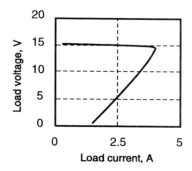

**Figure 5.7** *Output characteristic of the simulated regulator*

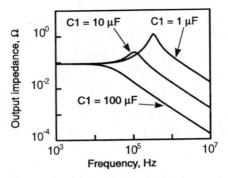

**Figure 5.8** *Output impedance of the simulated regulator*

large, the output impedance is free of peaks and is simply the impedance of C1 at very high frequency.

The stabilization factor of this regulator is not particularly good. Performing a small-signal (a.c.) analysis gives an output voltage of 10 mV with a voltage of 1 V injected at the input. This corresponds to a stabilization factor $S$ of 0.013. The main reason for this relatively large value is the slope resistance of the Zener voltage reference diode. A lower slope resistance, or a current source device to supply the current for the reference diode, would improve the performance significantly.

## 5.7 Over-voltage protection

Perhaps the most common fault to occur in a series regulator is the failure of the pass transistor, and it is probable that it will fail to short circuit. The effect is to subject the load to an over-voltage. To avoid a failure in the power supply destroying the load it is common to provide over-voltage protection. This is achieved by the use of a *crowbar*, which is a fast-acting, high-current switch which shorts the output of the regulator to ground. A thyristor is the usual choice of switch since thyristors are robust, have a high surge current rating and, once triggered, will remain in conduction until the current is turned off. To clear the fault current a fast-acting fuse is used in series with the regulator at either the source or load side. The thyristor must be triggered if an over-voltage occurs, and ICs are available for this. A simple circuit for over-voltage protection is shown in Figure 5.9.

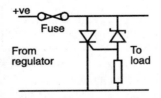

**Figure 5.9** *Simple method of providing over-voltage limiting*

## Summary

Linear regulators are used to obtain a constant voltage or current source from a supply that may vary with load current or time. They may be divided into two groups: shunt regulators, which may be simple, but are relatively inefficient, and series regulators. Protection in the form of over-current limiting and over-voltage protection using crowbars are essential for practical regulators to minimize damage under fault conditions.

## Self-assessment questions

5.1   A shunt regulator is to be used to provide a load voltage of 15 V. The source voltage may vary from 20 to 30 V and the load current from 0 to 100 mA. Calculate the maximum value for the series resistance, assuming that the current in the shunt regulator is not less than 25 mA. What will be the maximum power dissipation in (a) the resistance and (b) the shunt regulator? You may neglect the slope resistance of the shunt regulator.

5.2   A simple shunt regulator uses a 4.7 V Zener diode with a slope resistance of 10 $\Omega$. The supply voltage is 12 V and the series resistance is 470 $\Omega$. Calculate the stabilization factor and the output resistance.

5.3   For the regulator described in Question 5.2 if the input voltage is in the range 10–15 V and the load current in the range 0–5 mA, estimate the line regulation with no load current and the load regulation with a source voltage of 10 V.

5.4   Consider the series regulator simulated in Section 5.6. Estimate (a) the maximum power dissipated in the pass transistor while the regulator is in its normal unlimited mode of operation and (b) the power dissipation when the output is subjected to a short circuit.

## Tutorial questions

**5.1**   Why is a shunt regulator usually less efficient than a series regulator when the supply is essentially a voltage source?

**5.2**   If the power source were an ideal current source what type of regulator would be appropriate?

**5.3**   Explain why a series regulator used to produce a constant voltage requires three terminals, while to produce a constant current only two are required.

**5.4**   A manufacturer wishes to specify the performance for the regulator to be used in the power supply of equipment. Consider the relative merits of the stabilization factor and the line regulation as ways to specify the required stability against variations of source voltage.

**5.5**  Why is over-current protection an almost universal feature of series regulators?

**5.6**  Why do variable bench power supplies usually not feature foldback current limiting?

**5.7**  In what sort of equipment would over-voltage protection usually be required?

CHAPTER 6

# *Control of power by switching*

## 6.1 Introduction

Linear regulators, as discussed in the previous chapter, may provide good stable voltage (or current) supplies, but they are inherently inefficient. The supply voltage must exceed the load voltage and the supply current must exceed the load current. A significant fraction of the total input power will be dissipated in the active device(s) which control the load voltage (or current). These limitations may be overcome by the use of switching techniques. The active devices are used as power switches. When turned off they obviously dissipate little power as no current flows, while when turned on the voltage across the device is small and so again little power is dissipated in the device. In order to provide continuously variable power to the load, the switches are turned on and off rapidly, with the mean power delivered being controlled by the fraction of the time for which the switch is on (the *duty ratio*). If the period of this switching cycle is short compared with the time constant of the load, then essentially the power supplied to the load may be varied continuously by varying the duty ratio. If the switching period is constant and the on-time is varied this is referred to as pulse-width modulation (PWM). Alternatively, if the on-time is constant and the period varied this is pulse frequency modulation (PFM).

In some applications the pulsed current in the load may present no problems (for example, a resistive heater element where the thermal time constant is long compared with the switching period). However, if the load current must be continuous, with only residual ripple, this may be achieved by including inductance in series with the load. In some applications the load may already present a significant inductance (for example, the windings of a d.c. motor). With an inductive load the load current cannot simply be turned on and off as the voltage is switched, the current can only ramp up or down. In this chapter the use of transistors to switch the current in inductive loads is described, together with consideration of the losses incurred in the transistor during switching and the protection of the switch from excessive voltage or power dissipation.

## 6.2 Switching current in an inductive load

If the load is inductive, the current in the load cannot simply be turned on or off by a series switching device. The voltage across an inductor may change

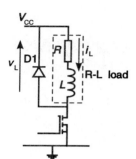

**Figure 6.1** *An inductive load switched by a series transistor with a freewheeling diode*

discontinuously, but it is the rate of change of current that is controlled by the voltage across the inductor. When switching current in an inductive load the current can only be diverted by the switch to an alternative path and the current in the inductance will continue to flow. This alternative path is most often obtained by the provision of a *freewheeling diode* connected across the load as illustrated in Figure 6.1. This transfer of current from one device to another is referred to as *commutation.* In order to understand the operation of many power circuits it is essential to grasp the concept that when switching inductive circuits the current can only be switched from one path to another, not suddenly interrupted.

When the transistor in Figure 6.1 is on, the current in the load ($i_L$) will build up at a rate determined by

$$L\frac{di_L}{dt} + Ri_L = V_{CC} \tag{6.1}$$

where the voltage drop across the transistor has been neglected. When the transistor is turned off the current must continue to flow in the inductor. As the switch current falls, the voltage across the inductor will rise rapidly, as the inductor current charges the stray capacitance. When the voltage has risen sufficiently, the freewheeling diode will be forward biased, and the current will be commutated, or diverted, from the switch to the freewheeling diode (D1). The current in the inductor will then decay at a rate given by

$$L\frac{di_L}{dt} = -Ri_L \tag{6.2}$$

where the voltage drop across the diode has been neglected.

If the current in the load falls to zero before the start of the next pulse, then the current in the load will be *discontinuous,* if not, then the load current will be *continuous.* Note that for a simple inductive–resistive load the current will always be continuous if the diode forward voltage is ignored, since the current will rise and fall exponentially. Although, of course, if the time for which the switch is off greatly exceeds the time constant ($L/R$) the current will fall to a very small value.

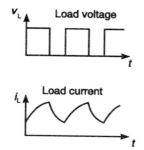

**Figure 6.2** *The load voltage and load current in an inductive load with a switched voltage*

The voltage and current waveforms across the load are shown in Figure 6.2, neglecting the diode forward voltage and the voltage across the transistor while it is on. The mean load voltage, $V_L$, is given by

$$V_L = \frac{V_{CC}t_{on}}{T} = V_{CC}D \tag{6.3}$$

where $t_{on}$ is the time for which the transistor is on, $T$ is the switching period and $D$ is the duty ratio ($t_{on}/T$). Since the mean voltage across the inductor must be zero, the mean load current, $I_L$ is given by

$$I_L = \frac{V_L}{R} \tag{6.4}$$

When switching the current in a highly inductive load the time taken for the switch to turn on or off will generally be sufficiently short that the load current may be taken as constant during the transition. During the switch transition the current must be diverted to or from the freewheeling diode. For this transition to be possible the diode must be forward biased, so while the current is transferred from the switch to the diode, or the diode to the switch, the load voltage must be small and negative while the transistor voltage must equal the supply voltage, $V_{CC}$, plus the diode forward voltage. A simplified diagram illustrating the current and voltage waveforms is shown in Figure 6.3. As the switch turns on the switch current rises, and, at the same time, the diode current falls. The load voltage does not start to rise until all the load current has been transferred to the switch. Similarly, at turn-off the voltage must rise as soon as the switch current starts to fall.

### EXAMPLE 6.1

A load with a resistance of 10 $\Omega$ and an inductance of 1 mH is driven from a d.c. voltage source of 100 V with pulse-width modulation using the circuit shown in Figure 6.1. The switching frequency is 25 kHz and the duty cycle 0.25. Explain why the current in the load must be continuous. Deduce the mean value of the load current and the mean power dissipated in the load.

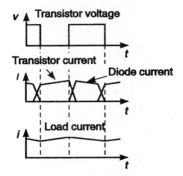

**Figure 6.3** *The load voltage, load current and the currents in the transistor and freewheeling diode with a highly inductive load*

## SOLUTION

The time constant of the load, $L/R$, is 100 $\mu$s, while the period of the switching is 40 $\mu$s (10 $\mu$s on and 30 $\mu$s off). Since the off-time is much less than the time constant the current cannot decay to zero in the off-time and the current will be continuous.

The mean load voltage = $0.25 \times 100 = 25$ V.

The mean load current = $25/10 = 2.5$ A.

The mean load power = $(2.5)^2 \times 10 = 62.5$ W.

## EXAMPLE 6.2

For the system described in Example 6.1, calculate the peak-to-peak ripple current in the load.

## SOLUTION

(a) With the switch on the current will obey the equation

$$0.001\frac{di}{dt} + 10i = 100$$

which has a solution of the form $i = 10 + A \exp(-10^4 t)$. If when the switch turns on at time $t = 0$ the current is $i_1$ then, $A = i_1 - 10$, hence, while the switch is *on*, $i = 10 + (i_1 - 10) \exp(-10^4 t)$. The switch turns off after 10 $\mu$s when the current is given by $i_2 = 0.9516 + 0.9048 i_1$.

(b) When the switch is off the load current decays exponentially with a time constant $10^{-4}$ s for 30 $\mu$s. If $i_3$ is the current at the end of the off-time then $i_3$ is given by

$$i_3 = i_2 \exp(-10^4 \times 30 \times 10^{-6}) = 0.7408 i_2$$

Clearly in the steady state $i_3$ must equal the current at the start of the cycle $i_1$. Hence equating $i_3$ and $i_1$ and solving for $i_1$ and $i_2$ gives 2.138 A and 2.886 A. The peak-to-peak ripple will be $2.8886 - 2.138 = 0.748$ A.

**ALTERNATIVE APPROXIMATE SOLUTION**

A simpler, approximate method of solution is possible since the time constant (100 $\mu$s) is much greater than the time for which the switch is off (30 $\mu$s). In this case the ripple current will be much less than the mean load current. We can therefore replace the load current $i_L$ in equation (6.2) by the mean load current $I_L$ which may then be integrated to give

$$\Delta i_L \approx RI_L t_{off}/L$$

where $t_{off}$ is the time for which the switch is off. Substituting values of $10\Omega$ for $R$, 2.5 A for $I_L$ , 30 $\mu$s for $t_{off}$ and 1 mH for $L$ gives a peak-to-peak ripple current of 0.75 A, in very close agreement with the exact solution.

## 6.3  Losses in the switching transistor

### 6.3.1  On-state losses

There are two distinct loss mechanisms to consider. If the transistor is on then there will be power dissipated due to the finite voltage across the transistor switch. For a MOSFET the voltage drop across the transistor is characterized by its on-state resistance $R_{DS(on)}$ and the associated power dissipation in the transistor ($P_{on}$) is given by

$$P_{on} = R_{DS(on)} < i_T^2 >$$

where $i_T$ is the transistor current and the brackets indicate a time average. If the ripple current in the load is small compared with the mean load current the power loss may be approximated by

$$P_{on} = R_{DS(on)}I_L^2 D \tag{6.5}$$

where $I_L$ is the mean load current and $D$ the duty ratio. For low-voltage power MOSFETs $R_{DS(on)}$ may be very low and on-state losses are frequently small.

### 6.3.2  Switching losses with an inductive load

The second source of power loss is that incurred when the transistor turns on (or off). When switching an inductive load as described in the previous section the switch will be subjected to the full supply voltage while the current in the switch rises and falls. Thus the instantaneous power will reach a maximum given by the product of the load current at the time of switching and the supply voltage. In order to calculate the average power loss during switching it is convenient, and reasonably realistic, to assume that the current in the transistor switch rises and falls linearly with time, and that the switch-on state voltage and the diode forward voltage may be ignored.

The switch current during turn-on is given by

$$i = \frac{I_L t}{t_r}$$

where $I_L$ is the load current (assumed constant) and $t_r$ is the rise time of the current in the switch. The total energy loss during a switch transition is found by integrating the instantaneous power dissipation, $i\,V_{CC}$, over the time taken for the transition. At turn-on the energy loss, $E_{on}$ is given by

$$E_{on} = \frac{V_{CC} I_L}{t_r} \int_0^{t_r} t\,dt = \frac{I_L V_{CC} t_r}{2} \qquad (6.6)$$

Similarly, at turn-off the energy lost, $E_{off}$ is given by

$$E_{off} = V_{CC} I_L \int_0^{t_f} \left(1 - \frac{t}{t_f}\right) dt = \frac{V_{CC} I_L t_f}{2} \qquad (6.7)$$

where $t_f$ is the fall time for the current in the switch. Thus the average power dissipation due to the switching losses, $P_{SW}$, is given by

$$P_{SW} = (E_{on} + E_{off})f = \frac{V_{CC} I_L (t_r + t_f)f}{2} \qquad (6.8)$$

where $f$ is the switching frequency.


### 6.3.3 Losses due to reverse recovery of the freewheeling diode

The discussion above has assumed that the diode turns on and off instantaneously as the current is reversed. This is not generally the case. For junction diodes, as discussed in Chapter 1, when the diode has been carrying a forward current the stored minority charge must be removed before the diode can turn off. The effect is that as the switch turns on the current flowing in the diode is reduced, then reversed for a short interval. This may lead to a large transient current flowing through both the switch and the diode.

During the reverse recovery of the diode, the switch must supply the load current and the diode recovery current. Also, during this time the diode voltage will remain small at close to its on-state value. Thus the energy lost as the switch turns on will be increased by an amount, $E_{rr}$, given by

$$E_{rr} = \int_0^{t_{rr}} V_{CC}(I_L + i_{rr})dt$$

$$= V_{CC} I_L t_{rr} + V_{CC} Q_{rr} \qquad (6.9)$$

where $i_{rr}$ is the reverse recovery current, $t_{rr}$ is the recovery time and $Q_{rr}$ is the reverse recovery charge.

### 6.3.4 Switching losses with a resistive load

With a resistive load $R_L$ the voltage falls as the current rises, and the switching losses are reduced. The energy loss at turn-on is given by

$$E_{on} = \int_0^{t_r} (V_{CC} - iR_L)i\,dt \qquad (6.10)$$

where, assuming a linear rise of switch current, $i$ is given by

$$i = \frac{V_{CC}t}{R_L t_r}$$

Integrating equation (6.10) gives

$$E_{on} = \frac{V_{CC}^2 t_r}{6R_L} = \frac{V_{CC}I_L t_r}{6} \qquad (6.11)$$

where $I_L$ is the load current $V_{CC}/R_L$.

A similar expression is obtained for $E_{off}$, hence the average power dissipation due to the switching losses is given by

$$P_{SW} = \frac{V_{CC}I_L(t_r + t_f)f}{6} \qquad (6.12)$$

or one-third of that when switching an inductive load with the same load current.

### EXAMPLE 6.3

A PWM controller using a MOSFET switch supplies current to an inductive load. The d.c. supply has a voltage of 50 V and the mean load current is 5 A, with negligible ripple. The switching frequency is 100 kHz and the duty ratio 0.7. The transistor has an $R_{DS(on)}$ of 0.1 $\Omega$ and the current rise and fall times are 0.2 $\mu$s. The freewheeling diode has a forward voltage drop of 0.8 V at 5 A, and its reverse recovery time may be neglected. Estimate (a) the mean power dissipated in the switching transistor and (b) the overall efficiency of the controller.

### SOLUTION

(a) On-state losses are given by

$$P_{on} = R_{DS(on)}I_L^2 D = 0.1 \times 5^2 \times 0.7 = 1.75 \text{ W}$$

while the switching losses are given by

$$P_{SW} = \frac{V_{CC}I_L(t_r + t_f)f}{2} = \frac{50 \times 5 \times (2 \times 10^{-7} + 2 \times 10^{-7}) \times 10^5}{2}$$

$$= 5 \text{ W}$$

Thus the total power loss in the transistor is 6.75 W.

(b) Power loss in diode during freewheeling is $V_D \times I_L = 0.8 \times 5 \times 0.3 = 1.2$ W. Therefore the total power loss is $6.75 + 1.2 = 7.95$ W. The power delivered to the load is given by $I_L\ V_L$. The mean load voltage, $V_L$, is given by $(50 - 0.1 \times I_L) \times 0.7 + (-0.8) \times 0.3$, which is the average of the voltage across the load when the transistor is on and when it is off. The mean load voltage is therefore 34.4 V, and the mean load power $5 \times 34.4 = 172$ W, while the total power supplied is $172 + 7.95 = 180$ W. The overall efficiency is therefore given by $172/180 \times 100\% = 95.6\%$.

## 6.4  Switching trajectory

Section 6.3 considered the mean power dissipated within the switching devices. This is generally the main concern when considering overall efficiency or the size of heatsink required. The peak power dissipation may also be significant if the duration of the power pulse is sufficient to permit a significant rise in the junction temperature of the switch. A convenient method of graphical analysis is to plot on the safe operating area diagram for the switching device the locus of voltage and current as the switch turns on or off. This locus is the *switching trajectory*. If the switching trajectory lies within the safe operating area then the peak power is not excessive.

For an inductive load with a freewheeling diode the locus is very simple if the reverse recovery transient of the freewheeling diode is neglected. Figure 6.4 shows the shape of the trajectory when the current in the load is continuous. At turn-on, while the current in the switch rises to the value in the load, the load voltage remains zero (neglecting the freewheeling diode forward voltage) and the switch voltage remains at the supply voltage $V_{CC}$. When the transfer of the load current from freewheeling diode to switch is completed the switch voltage falls to a low

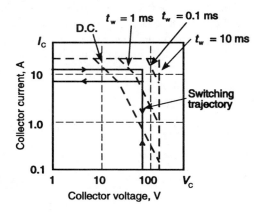

**Figure 6.4** *The switching trajectory for an inductive load with a freewheeling diode, superimposed upon the safe operating area of a bipolar transistor*

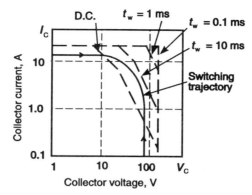

**Figure 6.5** *The switching trajectory for a resistive load, superimposed upon the safe operating area of a bipolar transistor*

value. At turn-off the voltage rises rapidly to $V_{CC}$ then the switch current decreases to zero. For the example shown in Figure 6.4 the switching trajectory will be comfortably inside the safe operating area provided the switching transition is not longer than 0.1 ms.

For a resistive load the peak power dissipation will be less than for an inductive load (with the same peak current and supply voltage), since the voltage falls as the current rises. On a linear plot the falling current with voltage would give a linear load-line, on the log-log plot of the safe operating area diagram the trajectory is curved as shown in Figure 6.5.

## 6.5 Switching aids or snubbers

Snubbers are circuits that limit the rate of rise of voltage across the switch and/or the rate of rise of current so as to modify the switching trajectory, limit the peak power dissipation and protect the switch against excessive rate of rise of current or voltage. Their function is therefore to limit the stress on the switch and aid the switching process. There are snubber circuits for use in a variety of circumstances but only two of the simplest circuits are considered here.

### 6.5.1 Turn-off snubber

The turn-off snubber limits the rate of rise of the voltage across a transistor switch driving an inductive load. The basic circuit is shown in Figure 6.6. When the current in the switching transistor starts to fall the switch voltage starts to rise and the load current is transferred through the diode D2 to the capacitor C. The capacitor charges up at an increasing rate as the switch current falls and more current is diverted to the capacitor. Assuming the switch current falls linearly, the current in the capacitor rises linearly, hence it is easy to show that the voltage

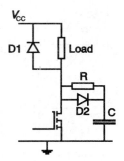

**Figure 6.6**  *A turn-off snubber for a transistor switch*

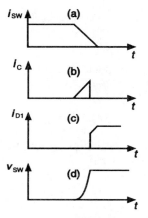

**Figure 6.7**  *Waveforms for a turn-off snubber. (a) Switch current; (b) capacitor current; (c) freewheeling diode current; (d) switch current*

across the capacitor rises quadratically. Neglecting the forward voltage of D2, the switch voltage will rise quadratically until either it reaches $V_{CC}$, when the current will be transferred to the diode D2, or the switch current falls to zero, in which case the switch voltage will continue to rise linearly until it reaches $V_{CC}$. Figure 6.7 shows the waveforms for the switch current, capacitor current freewheeling diode current and switch voltage for the case when the capacitor charges fully before the switch current falls to zero.

The energy stored in the capacitor is dissipated in the resistor when the switch turns on. The CR time constant must be short enough to allow the capacitor to fully discharge during the *on* time. However, the resistor must be large enough to limit the discharge current to a safe value when the transistor turns *on*.

With this simple circuit both the peak and mean power dissipation in the switch are reduced. However, the energy stored in the capacitor is all dissipated in the resistor. More complex circuits are able to recover the energy stored in the capacitor, improving efficiency.

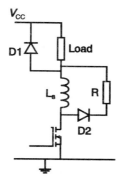

**Figure 6.8** *A turn-on snubber for a transistor switch*

### 6.5.2 Turn-on snubber

A simple turn-on snubber is obtained by including a small inductance in series with the switch. The rising current during turn-on induces a voltage across the inductance thereby reducing the switch voltage. If the snubber inductance is large enough the switch voltage will fall to zero while the current rises, and the rate of rise of current in the switch will be determined by the snubber inductance and the supply voltage $V_{CC}$. To discharge the energy stored in the snubber inductance a diode and resistor must be connected across the inductor as shown in Figure 6.8. The peak voltage across the switch at turn-off will rise to $V_{CC} + I_L R$ as the current in $L_S$ is diverted through $R$. The value of the resistance must be large enough to discharge $L_S$ while the switch is off, while avoiding excessive switch voltage or power dissipation at turn-off.

## 6.6 Simulation of switching with an inductive load

The use of simulation to study circuits for controlling power using PWM with a high switching frequency is complicated by a number of factors. The modelling of the circuit may be difficult because the behaviour during switch transitions may be greatly influenced by stray inductance and capacitance, which is in practice distributed rather than lumped into discrete components. There is also a very wide range of timescales, from tens of nanoseconds when considering the switch transitions to milliseconds or more when considering the transient behaviour of the load.

Consider a simple PWM controller with a MOSFET switch and an inductive load as shown in Figure 6.9. This is modelled in a simple form by the netlist given in Appendix 1 in Section A1.4.

The result of running this simulation is shown in Figure 6.10. The initial conditions are set by SPICE on the basis of a d.c. analysis, in this case with no

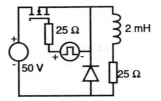

**Figure 6.9** *A circuit for the simulation of a switch with an inductive load*

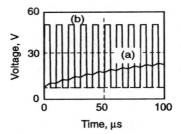

**Figure 6.10** *The simulated voltage (a) across the load and (b) across the resistive part of the load*

current in the load. The average voltage across the resistive part of the load is rising exponentially towards its steady-state value of 25 V, but after ten switching cycles has still not settled to its mean value. To examine the behaviour in the steady state either the simulation must be run for much longer or the initial conditions must be adjusted so that the simulation starts with the converter close to its steady state. This can be done by setting the initial voltages across capacitors and the initial currents in inductors. In this case it is quite easy since only the initial current in L1 needs be set to about 0.95 A. This is done by modifying the definition of L1 in the listing (see Appendix 1, Section A1.4) to:

```
L1  4  5  2mH  (IC = 0.95)
```

and the transient statement by adding (UIC) to force the use of initial conditions:

```
.tran  100n  10u  UIC
```

Making these modifications and running the simulation again gives the waveforms in Figure 6.11. The choice of suitable initial conditions is not always as straightforward, but it does enable the simulation to settle quickly. Hence the simulation takes less time, and it is easier to study details of the switching transitions, where it will generally be necessary to force the simulator to limit its maximum size for the time-step.

The diode model used in the above simulation did not include a value for the transit time, and hence has no reverse recovery transient. This may be changed by including the parameter Tt in the model parameter list. For a fast *p–n* junction diode

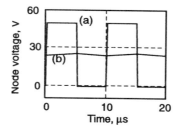

**Figure 6.11**  *The simulated voltage (a) across the load and (b) across the resistive part of the load with an initial current in the inductance*

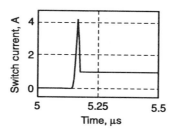

**Figure 6.12**  *The simulated current in the MOSFET showing the effect of reverse recovery in the freewheeling diode*

a value of about 50 ns is appropriate. Including this parameter gives a large spike in the MOSFET drain current (switch current), since both diode and MOSFET conduct simultaneously (Figure 6.12). Where the reverse recovery of the freewheeling diode leads to large current spikes, the only solution is to limit the rate at which the current is commutated from the switch to the diode by the use of a turn-on snubber.

## Summary

Pulse-width modulation of the load voltage is a common method of controlling power delivered to a load. The loads involved are frequently inductive, and although the voltage may be switched *on* and *off* by a series switch, the load current will only be transferred between the switch and an alternative path, usually a freewheeling diode. This concept of commutation of the current is fundamental to understanding many power electronic circuits. A transistor switching the current in an inductive load with a freewheeling diode will be subjected simultaneously to the maximum voltage and current, which may lead to significant losses of energy each time the switch is opened or closed. The power dissipation in the switch when switching an inductive load may be reduced by the use of snubbers which limit the rate of rise of voltage across the switch, or current through the switch and assist in the process of commutation between the switch and the freewheeling diode.

## Self-assessment questions

6.1 A load consists of an inductance of 10 mH and a resistance of 10 $\Omega$ in series. The load is energized from a source of 100 V using a series switch and a freewheeling diode to switch the load voltage with a period of 20 $\mu$s and a duty ratio of 0.25. Assuming that both switch and diode are ideal, what will be the average load current?

6.2 For the load and source described in Question 6.1, what will be the peak-to-peak ripple current? *Hint:* consider whether the ripple current will be a small fraction of the mean load current.

6.3 For the load and source described in Question 6. the pulse frequency is reduced to 1 kHz, while the duty ratio is kept constant. Will the mean load current have changed? What will be the new value of the peak-to-peak ripple current?

6.4 A bipolar transistor chops the voltage across a highly inductive load, with the current being transferred to a very fast freewheeling diode. If the source voltage is 200 V, the chopping frequency 50 kHz and the load current is almost constant with a value of 1 A, what will be the switching power loss in the transistor if it has a current rise time of 0.3 $\mu$s and a current fall time of 0.5 $\mu$s?

6.5 For the pulse-width modulator described in Question 6.5, what will be the total power loss in the switch and freewheeling diode if the duty ratio is 0.4, the saturation voltage of the transistor is 1 V and the diode forward voltage is 0.8 V?

6.6 A PWM controller with a turn-on snubber as shown in Figure 6.9 has a snubber inductance $L_s$ of 5 $\mu$H and a resistance $R$ of 10 $\Omega$. If the source voltage $V_{CC}$ is 50 V, what will be the maximum rate of rise of switch current, and what must the switch off-time be if the energy stored in the inductor is to be discharged before the switch turns on again?

6.7 For the PWM circuit with turn-on snubber described in Question 6.6, the switching frequency is 100 kHz and average load current is 2 A, with a peak-to-peak ripple of about 0.5 A. What will be the maximum switch voltage, and the power dissipated in the resistor $R$?

## Tutorial questions

**6.1** A load consists of an inductance in series with a resistance. The load is energized using PWM from a source of 100 V. A series switch and a free-wheeling diode switch the load voltage with a period of 20 $\mu$s and a duty ratio of 0.25. Explain why the load current, in this ideal case, will always be continuous.

**6.2** Explain the role of the freewheeling diode when switching current in an inductive load.

**6.3** Explain the origin of switching losses in a transistor switching current in an inductive load.

**6.4** Why does reverse recovery in a freewheeling diode increase the power loss if the load current is continuous? What effect will it have if the load current is discontinuous?

**6.5** Explain how a turn-off snubber reduces the energy dissipated in the switch when switching an inductive load.

**6.6** How will a turn-on snubber modify the switching trajectory when switching an inductive load with continuous current?

## References

Rashid, M. H., *Power Electronics: Circuits, Devices, and Applications*, 2nd edition, Prentice Hall, Englewood Cliffs, NJ, 1993.

Williams, B. W., *Power Electronics: Devices, Drivers and Applications*, Macmillan, London, 1987.

# CHAPTER 7
# *Switching converters*

## 7.1 Introduction

Switching converters transform d.c. from the potential of the source to another load potential which may be higher or lower than that of the source. They use switching techniques to achieve high efficiency and generally pulse-width modulation (PWM) to control the load voltage. Less frequently, pulse-frequency modulation (PFM) may be used. Together with an associated control circuit, a switching converter forms the basis of a switching regulator. With their high efficiency and small size, switching regulators have to a large extent replaced linear regulators in many types of electronic equipment, except where power is low and efficiency is not important, or where low electrical noise and high stability are required. This chapter aims to give an overview of the basic circuit topologies used in switching converters and to show how they operate.

The final part of the chapter introduces the concept of the transfer function for a converter. While this section minimizes the use of mathematics, it does assume some knowledge of transfer functions and the use of complex frequency. The results of this section are used in Chapter 8 to investigate the stability of the feedback loop in two design case studies. This section and the relevant parts of Chapter 8 may be omitted if the complexities associated with feedback control are not to be considered.

## 7.2 The buck converter or 'down' converter

The buck converter, or down converter, shown in Figure 7.1 is possibly the simplest switching regulator to understand. In essence, it is very similar to the basic switching controller with an inductive load considered in Chapter 6, except that in this case the inductor is part of the converter and there is a capacitor across the load. The series switching transistor S is turned on and off with a switching period $T$ and duty ratio $D$. The freewheeling diode D1 provides the alternative path for the current in the inductor $L$ when the transistor is turned off. The capacitor $C$ across the load ensures that the load voltage is essentially constant. The buck converter always gives a load voltage of less than the supply voltage.

When S is on, the current, $i_L$, in the inductor is given by

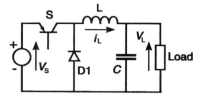

**Figure 7.1** *The buck converter*

$$L\frac{di_L}{dt} = V_S - V_L \qquad (7.1)$$

where the source and load voltages, $V_S$ and $V_L$, may be considered constants, and the switch-on-state voltage drop has been neglected. The solution to equation (7.1) is therefore

$$L(i_L - I_0) = (V_S - V_L)t \qquad (7.2)$$

where $I_O$ is the current when the switch turns on at time $t = 0$.

When the switch turns off then, neglecting the diode voltage drop, the current will decrease at a rate given by

$$\frac{di_L}{dt} = -\frac{V_L}{L} \qquad (7.3)$$

where $V_L$ is the mean load voltage (considered constant) and $i_L$ is greater than zero.

The current in the inductor ramps up linearly while the switch is on and down while the switch is off. If the inductor current has not decreased to zero by the end of the cycle then the converter is operating in *continuous current mode*, otherwise it is operating in *discontinuous mode*.

### 7.2.1 Continuous current

The switch is on for a time $DT$, so that while the switch is on the current increases by $(V_S-V_L)DT$. The switch is off for a time $(1-D)T$ and during this time the current will decrease by $-(1-D)TV_L$, assuming that the current does not fall to zero before the end of the cycle. In the steady state the current at the end of the cycle must be the same as the current at the start of the cycle, hence

$$(V_S - V_L)DT = V_L(1 - D)T$$

or

$$V_L = DV_S \qquad (7.4)$$

Provided the inductor current remains continuous the load voltage is independent of load current. The peak-to-peak ripple current in the inductor ($I_R$) is given by

$$I_R = \frac{(1 - D)T}{L}V_L \qquad (7.5)$$

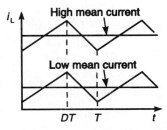

**Figure 7.2** *The inductor current in a buck converter in continuous current mode*

and is independent of the mean load current. The waveform of the inductor current is shown in Figure 7.2 for two values of the mean current. If the mean current falls below half the peak-to-peak ripple current then the inductor current will fall to zero and the current will become discontinuous.

### 7.2.2 Discontinuous current

In this mode the current builds up as given by equation (7.2), with the initial current $I_0$ equal to zero. The maximum inductor current when the switch turns off at time $DT$ is given by

$$I_m = \frac{V_S - V_L}{L} DT \qquad (7.6)$$

The time taken for the current to ramp down to zero ($T_0$) will be

$$T_0 = \frac{LI_m}{V_L} = \frac{V_S - V_L}{V_L} DT \qquad (7.7)$$

The inductor current waveform is shown in Figure 7.3. The total charge delivered to the capacitor in each cycle ($Q$) is the integral of the current over time $T$, which is just the area under the triangle of height $I_m$ and of base $DT + T_0$. Hence, the mean load current is given by

$$I_0 = \frac{Q}{T} = \frac{(DT + T_0)I_m}{2T} \qquad (7.8)$$

Substituting in equation (7.8) for $T_0$ from equation (7.7) and for $I_m$ from equation (7.7) gives, after some rearrangement,

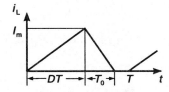

**Figure 7.3** *The inductor current in a buck converter in discontinuous current mode*

$$V_L = \frac{V_S}{1 + (2LI_0)/(D^2 TV_S)} \tag{7.9}$$

At zero mean load current the load voltage is clearly equal to $V_S$ and this decreases to $DV_S$ as the current increases to that necessary for continuous inductor current.

### EXAMPLE 7.1

A buck converter has an input voltage of 20 V and a load voltage of 5 V. The switching frequency is 100 kHz and the series inductor has a value of 300 $\mu$H. Neglecting losses in the series switch, the freewheeling diode and the inductor, estimate (a) the duty ratio and (b) the peak-to-peak ripple current if the load current is continuous, and (c) estimate the minimum load current for continuous conduction.

### SOLUTION

(a) The duty ratio $= V_L/V_S = 5/20 = 0.25$.

(b) The on-time is 2.5 $\mu$s and the off-time 7.5 $\mu$s. While the switch is off the current ramps down at a rate given by

$$\frac{di}{dt} = -\frac{V_L}{L} = -\frac{5}{3 \times 10^{-4}} = -16.67 \times 10^3 \text{ As}^{-1}$$

In 7.5 $\mu$s it will fall by 0.125 A, hence the peak-to-peak ripple current will be 0.125 A.

(c) The current will be continuous if the mean load current exceeds one half of the peak-to-peak ripple current, i.e. the mean load current exceeds 0.0625 A.

### EXAMPLE 7.2

For the buck converter described in Example 7.1, estimate the duty ratio if the freewheeling diode has a forward voltage of 0.8 V (assumed constant) while it is in conduction. Again, assume that the inductor current is continuous.

### SOLUTION

While the switch is off the current ramps down at a rate given by

$$\frac{di}{dt} = -\frac{V_L + V_D}{L} = -\frac{5.8}{3 \times 10^{-4}}$$

and while the switch is on it ramps up at a rate given by

$$\frac{di}{dt} = \frac{V_S - V_L}{L} = \frac{15}{3 \times 10^{-4}}$$

If $t$ is the on-time then $10^{-5} - t$ is the off-time, since the switching period is 10 $\mu$s. In equilibrium the current at the end of the cycle must be equal to that at its start, hence

$$t \times \frac{15}{3 \times 10^{-4}} = \frac{5.8 \times (10^{-5} - t)}{3 \times 10^{-4}}$$

from which $t$ is found to be 2.78 $\mu s$. Hence, the duty ratio will be 0.278.

## 7.3  The boost converter or 'up' converter

The buck converter always reduces the voltage, while, as its name suggests, the boost converter always steps up the voltage. The circuit of the basic converter is shown in Figure 7.4. In operation when the switch S is turned on, the current $i_L$ increases until the switch is turned off. Then the voltage across the switch increases rapidly, as the current continues to flow in the inductor. The voltage across the switch continues to rise until D1 is forward biased, and the current flowing in the inductor is diverted from the switch to the capacitor and the load. The polarity of the voltage across the inductor is now reversed, and the inductor current falls. As with the buck converter, discontinuous and continuous modes of operation are possible, depending on whether the inductor current falls to zero before the start of the next cycle.

### 7.3.1  Continuous current

While the switch is on the inductor current will rise linearly at a rate given by

$$\frac{di_L}{dt} = \frac{V_S}{L} \tag{7.10}$$

Hence, during the on-time, $DT$, the current will rise by

$$\Delta I_L = \frac{V_S DT}{L} \tag{7.11}$$

where $D$ is the duty ratio and $T$ the switching period. During the off-time the current will fall at a rate given by

$$\frac{di_L}{dt} = \frac{V_S - V_L}{L} \tag{7.12}$$

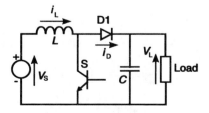

**Figure 7.4**  *The boost converter*

Since the duration of the off-time is $(1-D)T$ the current will fall by

$$-\Delta I_L = \left(\frac{V_S - V_L}{L}\right)(1 - D)T \qquad (7.13)$$

Eliminating $\Delta I_L$ from equations (7.11) and (7.13) the load voltage is given by

$$\frac{V_L}{V_S} = \frac{1}{1 - D} \qquad (7.14)$$

$D$ is always greater than zero and less than 1, hence the load voltage will exceed the supply voltage.

The currents in the inductor and diode are shown in Figure 7.5. The peak-to-peak ripple current, $I_R$, is equal to $\Delta I_L$ and is given by

$$I_R = \frac{V_S DT}{L}$$

### 7.3.2 Discontinuous current

If the mean source current falls below half the peak-to-peak ripple current calculated above, then the inductor current will become discontinuous. In this mode the inductor current increases by $\Delta I_L$ while the switch is on, starting from zero, hence the peak current will be given by

$$I_m = \Delta I_L = \frac{V_S DT}{L} \qquad (7.15)$$

While the switch is off, the current will decrease linearly to zero in a time

$$T_0 = \frac{L I_m}{V_L - V_S} = \frac{V_S DT}{V_L - V_S} \qquad (7.16)$$

The charge delivered to the capacitor each cycle is the integral of the diode current while the current falls to zero, that is,

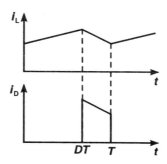

Figure 7.5  The inductor and diode currents in the boost converter operating in continuous current mode

$$Q = I_m T_0/2 = \frac{V_S^2 D^2 T^2}{2L(V_L - V_S)} \tag{7.17}$$

The mean load current is $Q/T$, hence equation (7.17) can be solved to give

$$\frac{V_L}{V_S} = 1 + \frac{V_S D^2 T}{2L I_0} \tag{7.18}$$

Clearly, $V_L$ may become very large as $I_o$ tends to zero. In practice, the maximum voltage will be limited by the breakdown of the switch, the diode or the capacitor. It is clearly desirable to prevent this by the use of a protection circuit to detect the excess voltage and stop the switching, and possibly an avalanche diode across the switch to protect it from accidental over-voltage.

The limit of the discontinuous current regime may be found by setting the value of the load voltage for discontinuous current equal to that for continuous current, i.e. setting $V_L/V_S$ equal to $(1-D)^{-1}$, and solving for $I_o$. This gives

$$I_0 = \frac{V_S(1 - D)DT}{2L} \tag{7.19}$$

Using equation (7.11) this condition may be related to the inductor peak-to-peak ripple current ($I_R$), to give the condition for continuous conduction in terms of the load current as

$$I_0 > \frac{(1 - D)I_R}{2} \tag{7.20}$$

### EXAMPLE 7.3
A boost converter transforms a source voltage of 10 V to a load voltage of 50 V. The switching frequency is 25 kHz, the inductance 100 $\mu$H and the load resistance 500 $\Omega$. Show that the current is discontinuous.

### SOLUTION
There are several methods that could be used to show that the current must be discontinuous. Possibly the simplest is to assume that it is *continuous* and show that equation (7.20) is *not* satisfied, hence the current must be discontinuous.

The transformation ratio ($V_L/V_S$) is 5, assuming continuous current $D$ is 0.8. The peak-to-peak ripple current is given by

$$I_R = \frac{V_S DT}{L} = \frac{10 \times 0.8 \times 40 \times 10^{-6}}{100 \times 10^{-6}} = 3.2$$

Hence the condition for continuous current becomes

$$I_0 > \frac{0.2 \times 3.2}{2} = 0.32 \text{ A}$$

Since the mean load current is only 0.1 A, this condition is clearly not satisfied and the current must be discontinuous.

### EXAMPLE 7.4
For the converter described in Example 7.3, what will be (a) the duty ratio and (b) the peak inductor current?

### SOLUTION
(a) The current is discontinuous, hence equation (7.18) may be used to calculate the duty ratio. The load current, $I_0$, is 50/500 = 0.1 A:

$$5 = 1 + \frac{10 \times D^2 \times 40 \times 10^{-6}}{2 \times 10^{-4} \times 0.1}$$

Hence the duty ratio $D$ is 0.4472.

(b) The peak inductor current is found from equation (7.15) and is given by

$$I_m = -\frac{10 \times 0.4472 \times 40 \times 10^{-6}}{10^{-4}} = 1.789 \text{ A}$$

Note that in discontinuous mode the peak current may be very much greater than the mean current.

## 7.4 The buck–boost converter or 'up–down' converter

The third of the basic switching converter circuits is the buck–boost circuit, which is able to step the voltage up or down. It also reverses the polarity of the supply. The configuration is as shown in Figure 7.6. While the switch is closed energy is stored in the inductor. The inductor current increases until the switch is opened. When the switch opens, the current in the inductor continues flowing. The voltage across the inductor is reversed as the current continues to flow and charges the stray capacitance until the diode is forward biased, and the inductor current is transferred to the load. As with the other types of converter, the inductor current may be continuous or discontinuous.

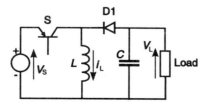

**Figure 7.6** *The buck–boost converter*

### 7.4.1 Continuous current

While the switch is on the inductor voltage equals the source voltage, so that during the on-time ($DT$) the inductor current will increase by

$$\Delta I_L = V_S \frac{DT}{L} \tag{7.21}$$

and while the switch is off the current will decrease by

$$-\Delta I_L = V_L \frac{(1-D)T}{L} \tag{7.22}$$

Assuming a steady state and eliminating $\Delta I_L$ from equations (7.21) and (7.22) gives the ratio of load to source voltage:

$$\frac{V_L}{V_S} = -\frac{D}{1-D} \tag{7.23}$$

Note that $V_L$ and $V_S$ are of opposite sign and the ratio may be greater or less than 1.

### 7.4.2 Discontinuous current

In discontinuous current mode the load voltage can be deduced as before by integrating the inductor current during discharge and relating this to the mean load current. An alternative approach is to use the stored energy. With this converter in discontinuous current mode, energy is stored in the inductor while the switch is on, and no energy is transferred to the load. When the switch is turned off, all the stored energy is transferred to the load.

While the switch is on, the inductor stores an energy given by

$$E = \frac{LI_m^2}{2} = \frac{L}{2}\left(\frac{V_S DT}{L}\right)^2 = \frac{V_S^2 D^2 T^2}{2L} \tag{7.24}$$

The mean load power is $-V_L I_0$, where $I_0$ is the mean load current, which is negative, hence the sign. Equating the mean load power to $E/T$ the rate of transfer of energy from the inductor gives

$$\frac{V_L}{V_S} = -\frac{V_S D^2 T}{2LI_0} \tag{7.25}$$

## 7.5 Isolation of load from source

The three converters described above all provide a direct electrical path between source and load along the common line. For many purposes isolation between source and load is essential. This can be achieved by modifying the basic converter topologies so as to include a transformer, which must, of course, operate at the switching frequency.

### 7.5.1 An isolated buck or forward converter

An example of a buck converter with an isolation transformer is shown in Figure 7.7. While the switch S is on, the supply voltage is applied across the primary of the transformer which in turn induces a voltage $nV_S$ across the secondary, where $1:n$ is the turns ratio of the transformer. The current induced by this e.m.f. flows through D2 to the inductance $L$ and the load. When the switch turns off the induced current stops and the inductor current is carried by the freewheeling diode D3 as in the simple buck converter. Also as for the simple buck converter, continuous and discontinuous modes of operation are possible. The load voltage can be made greater or less than the source voltage by a suitable choice of $n$. For this reason the name *forward converter* is often preferred.

When the switch turns off there is flux stored in the transformer core. This flux may be viewed as due to the magnetization current flowing in the primary inductance ($L_P$) of the transformer and an alternative path must be provided while it decays to zero. A freewheeling diode together with some series resistance ($R$) could be used across the primary to provide a path for the current and permit a negative voltage across the primary, thereby allowing the current in the primary inductance to decrease with a time constant $L_P/R$.

A better solution is shown in Figure 7.7. The third winding (the reset winding) and the diode D1 allow the energy stored in the magnetic flux to be returned to the power supply. This process is referred to as resetting the core of the transformer. During the resetting phase the voltage across the switch will exceed the power supply voltage, and a negative voltage will appear across the transformer secondary. D2 prevents this negative voltage from reaching the inductor and freewheeling diode.

The time taken to reset the core is determined by the turns ratio between primary and reset windings. If both have the same number of turns the reset time will equal the duration for which the power switch is on, and the peak voltage across the

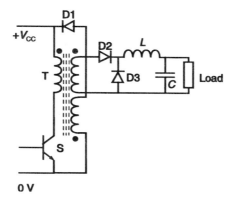

**Figure 7.7** *An isolated forward converter*

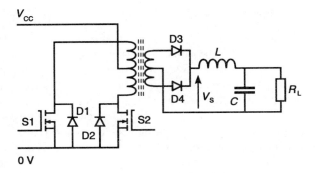

**Figure 7.8** *The push–pull, isolated forward converter*

switch will be twice the source voltage. The peak switch voltage may be reduced by increasing the number of turns on the reset winding, but this will increase the reset time. The need to reset the core, and the trade-off of reset time against maximum switch voltage, will limit the maximum duty ratio, usually to less than 0.5.

### 7.5.2 A push–pull forward converter

The forward converter described in the previous section can be quite efficient since the energy stored in the transformer core is returned to the power source. However, it must operate at a relatively low duty ratio. Also, the flux in the transformer core is limited to the range of zero up to the maximum working flux density $B_{max}$. If the transformer primary can operate in a bipolar manner then the flux density swing can be increased to $-B_{max}$ to $+B_{max}$, reducing the size and weight of the core, and avoiding the need for any special circuit to reset the flux in the core.

A bipolar drive is often achieved by using the push–pull configuration of Figure 7.8, with two power switches and a centre-tapped transformer. The switches are timed to be on during alternate halves of the cycle. Each switch must be on for less than half a cycle to avoid both switches conducting simultaneously, but the duty ratio as seen at the load side of D3 and D4 may approach unity. A separate freewheeling diode is not required as the current in $L$ will freewheel through D3 and D4. A simplified schematic of the currents and voltages is shown in Figure 7.9.

The current flowing through D3 and D4 during the freewheeling phase will also support the flux in the transformer core by splitting unequally, since, in general, the load current will be much greater than the magnetization current for the transformer. The reverse diodes D1 and D2 are necessary to carry the current due to the leakage flux of the transformer.

A disadvantage of the push–pull arrangement is that the peak voltage across the switching transistors is twice the source voltage. For an off-line switching power supply, the d.c. source may have a potential of around 380 V, which puts the voltage rating of the switches out of the range of the more common types of

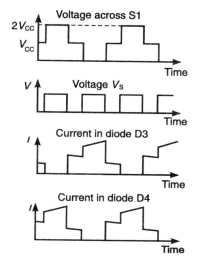

**Figure 7.9** *Voltages and currents in a push–pull forward converter*

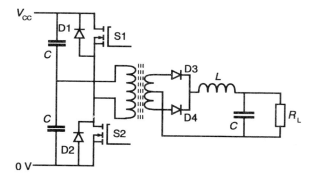

**Figure 7.10** *A half-bridge forward converter*

MOSFET. In this case a half-bridge configuration, as shown in Figure 7.10, may be preferred, since for this configuration the peak switch voltage is limited to the source voltage. Half- and full-bridge switching circuits will be considered in Chapter 10.

### 7.5.3 An isolated buck–boost or flyback converter

A popular isolated converter for use at low to moderate power levels is the isolated buck–boost or *flyback* converter. The basic circuit is shown in Figure 7.11. The transformer T combines the role of transformer and energy storage element. While the switch S is on, the current builds up in the transformer primary, storing energy. When the switch turns off the polarity of primary and secondary voltages reverse,

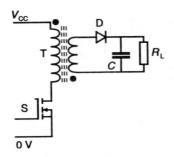

**Figure 7.11** *An isolated flyback converter*

the diode D starts to conduct and the stored energy is transferred to the reservoir capacitor C. The converter is normally used in a mode such that the flux in the core drops to zero while the switch is off. This is the equivalent of discontinuous current mode. However, the converter could be operated in a mode equivalent to continuous current mode, in which the flux in the core is always non-zero, although, of course, neither primary or secondary currents are actually continuous. A power supply based on this converter will be discussed in Chapter 8.

## 7.6 Control of the basic converters

### 7.6.1 Steady-state transfer characteristics

The steady-state (d.c.) transfer characteristics of the basic converters, i.e. the relationship between input voltage, duty ratio and load voltage, have been derived for the basic converter topologies in Sections 7.2 to 7.4. Generally, when these converters are used it will be necessary to include them in a feedback loop to control the load voltage. The duty ratio is varied to control the load voltage, usually keeping the switching frequency constant. Various strategies are used to control the duty ratio, and the simplest, and the only one to be considered here, is the direct method, in which a control voltage, $V_C$, is used to control the pulse width.

The most common form of pulse-width modulator operates by comparing the control voltage with the voltage from a constant-frequency oscillator which generates a sawtooth waveform. This is illustrated in the schematic diagram in Figure 7.12. If the voltage from the oscillator is less than the control voltage then the comparator output is high, otherwise it is low. The waveforms are illustrated in Figure 7.13. If the amplitude of the sawtooth waveform is $V_{st}$ then the duty ratio of the modulator is given by

$$D = \frac{V_C}{V_{st}} \tag{7.26}$$

Substituting from equation (7.26) into the appropriate transfer characteristic gives the mean load voltage in terms of the control voltage $V_C$. For example, taking

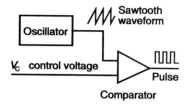

**Figure 7.12** *A simplified pulse-width modulator*

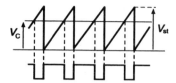

**Figure 7.13** *The waveforms for a pulse-width modulator*

the buck converter with continuous inductor current, the relation between the load voltage $V_L$ and the control voltage is

$$\frac{V_L}{V_C} = \frac{V_S}{V_{st}}$$ (7.27)

This relation gives the effective d.c. gain of the power stage and modulator. In this case the effective gain depends only on the source voltage and the amplitude of the sawtooth voltage used in the pulse-width modulator. However, in general, this will be a non-linear relationship, and the effective gain will depend on the source voltage, the load voltage, and the load current.

Consider, for example, the buck–boost converter with discontinuous current (equation (7.25)). In this case the transfer function becomes

$$\frac{V_L}{V_C} = \left(\frac{V_S^2 T}{2LI_0}\right)\frac{V_C}{V_{st}^2}$$ (7.28)

This is clearly non-linear with respect to the control voltage $V_C$. In order to use such a relationship to design a feedback control system it is usual to linearize the equation by considering small deviations from the steady state.

### 7.6.2 Dynamic transfer characteristics

When a converter is used within a control loop it is essential to understand the dynamic behaviour of the power stage in order to ensure stability and proper dynamic behaviour of the feedback loop. One technique which has been widely used to produce a linear, small-signal, dynamic model of power conversion stages is *state-space averaging*. This technique of analysis is beyond the scope of this book. A good introduction to the use of state-space averaging is given by Mohan,

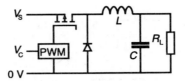

**Figure 7.14**  *A simple buck converter*

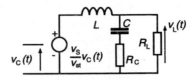

**Figure 7.15**  *The small-signal time-averaged model of a buck converter*

Undeland and Robbins (1995), but a more detailed description of the modelling of converter dynamics is given by Kislovski, Redl and Sokal (1991), or Kassakian, Schlecht and Verghese (1991). Here an *ad-hoc* approach is taken to analyze the buck converter in continuous conduction and the flyback converter in discontinuous conduction.

The buck converter shown in Figure 7.14, when operating in continuous conduction, can be modelled relatively simply. The small-signal model is used, in which a variable (a current or a voltage) is represented by the sum of its quiescent value and its small-signal value. The quiescent value is constant, while the small-signal value is a function of time. It is further assumed that the small-signal values change only slowly compared with the time scale represented by the switching period, and thus may be replaced by their average values. The averaging is performed over a complete switching cycle. The control voltage is represented by the sum of its quiescent value, $V_C$, and its small-signal average value, $v_C(t)$. The load voltage is represented by its quiescent value, $V_L$, plus its average value $v_L(t)$. The dynamic behaviour of the converter may then be modelled by the small-signal equivalent circuit shown in Figure 7.15. Note the inclusion of the equivalent series resistance of the capacitor, as this has a significant effect on the transfer function.

This simple model is obtained because the filter inductor, capacitor and load all remain connected in the same way, whether the power switch is open or closed. The only change in the equivalent circuit during the switching period is the cyclic change in the voltage at the input of the LC filter. The simple model of Figure 7.15 may be justified properly by the use of the state-space averaging technique. In discontinuous current mode this model does not work. Neither is it possible to model the other converter topologies in such a simple manner, so directly related to the original circuit.

Using the model shown in Figure 7.15, the transfer function of the converter may be derived in terms of the Laplace transforms of the small-signal load and control voltages, and is given by

$$H(s) = \frac{\tilde{v}_L(s)}{\tilde{v}_C(s)} = \frac{V_S}{V_{st}} \frac{1 + s/\omega_z}{1 + (s/\omega_0)/Q + (s/\omega_0)^2} \qquad (7.29)$$

where $\tilde{v}_L(s)$ and $\tilde{v}_C(s)$ are the Laplace transforms of the load and control voltages and

$$\omega_0 = \frac{1}{\sqrt{LC}}$$

$$\omega_z = \frac{1}{R_c C}$$

$$Q = \frac{R_L}{\omega_0 L}$$

The implications of this transfer function when applied to a practical example of a converter will be considered in Chapter 8.

### EXAMPLE 7.4
A buck converter operating in continuous current has a source voltage of 48 V and load voltage of 12 V with a load resistance of 12 Ω. The inductance has a value of 500 μH, the capacitor has a value of 470 μF, with an equivalent series resistance of 100 mΩ, and the switching frequency is 25 kHz. The pulse-width modulator has a sawtooth waveform of amplitude 3.5 V. Show that the current in the inductor will be continuous, and calculate the small-signal transfer function of the converter from control voltage to load.

### SOLUTION
The duty ratio will be 12/48 = 0.25, hence the switch will be off for a time of 0.75×40 μs = 30 μs. Hence, assuming continuous inductor current, the peak-to-peak ripple current in the inductor is given by $12 \times 30 \times 10^{-6}/(500 \times 10^{-6}) = 0.72$ A. The load current is 1 A which is more than half the peak-to-peak ripple current, hence, the assumption of continuous current is valid.

The transfer function is given by equation (7.29). Substituting in values for the various components gives $\omega_0 = 2.06\times10^3$, $\omega_z = 2.1\times10^4$ and $Q = 11.7$. The effective gain $V_S/V_{st} = 48/3.5 = 13.7$. Hence the transfer function is given by

$$H(s) = 13.7 \left( \frac{1 + 4.76 \times 10^{-5} \, s}{1 + 4.17 \times 10^{-5} \, s + 2.36 \times 10^{-7} \, s^2} \right)$$

For a converter operating in discontinuous mode the inductor current is not influenced by its value in the previous switching cycle, since it must fall to zero before the end of the cycle. The inductor current and reservoir capacitor cannot act as resonator, exchanging energy at a frequency below the switching frequency. This means that in discontinuous conduction the pair of complex poles observed for the buck converter in continuous conduction could not be obtained. While the buck converter is not normally used with discontinuous conduction in the inductor, the flyback converter is frequently used in this mode, and will therefore be analyzed.

For the flyback converter operating in discontinuous mode the energy stored in the inductor depends only on the duty ratio and the source voltage. All this energy is transferred to the load during the second half of the cycle. The charge transferred will depend on the load voltage. Assuming a load resistance $R_L$, the steady-state output current may be obtained from equation (7.25), and is given by

$$I_0 = V_S D \sqrt{\frac{1}{2LfR_L}} = \frac{V_S V_C}{V_{st}} \sqrt{\frac{1}{2LfR_L}} \qquad (7.30)$$

where $f$ is the switching frequency $(1/T)$ and the duty ratio has been expressed in terms of the control voltage, $V_C$, and the amplitude of the sawtooth waveform in the pulse-width modulator, $V_{st}$.

A small-signal model of the flyback converter, in which the voltages and currents are considered to be averaged over the switching period, can be produced by modelling the power transfer using a current source in parallel with a resistance. Both components may be non-linear. The circuit model is illustrated in Figure 7.16. The power supplied by the source and transferred to the load is equal to the energy stored in the inductor (from equation (7.24)) multiplied by the frequency, and is given by

$$P = \frac{V_S^2 D^2}{2Lf}$$

This is equivalent to an input resistance $R_{in}$ given by

$$R_{in} = \frac{2Lf}{D^2}$$

The value of the output resistance, $R_0$, is found by differentiating equation (7.25) with respect to the load current:

$$R_0 = -\frac{\partial V_L}{\partial I_0} = \frac{V_S D^2}{2Lf V_L^2} = \frac{V_L}{I_0} = R_L \qquad (7.31)$$

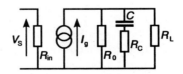

**Figure 7.16** *The time-averaged model of a flyback converter*

Since the output resistance equals the load resistance, the equivalent current generator must have a value, $I_g$, equal to twice that given by equation (7.30), in order to obtain the correct load current, i.e.

$$I_g = \frac{2V_SV_C}{V_{st}}\sqrt{\frac{1}{2LfR_L}}$$

The load current depends linearly on the control voltage, so no linearization is required, and since it only depends on the control voltage during the single switching period, there will be no dependence on previous values, and hence no frequency dependence, subject to the general requirement that the frequencies considered are much less than the switching frequency. Hence, the small-signal current source has a value, $i_t$, given by

$$i_t(t) = \frac{2V_S}{V_{st}}\sqrt{\frac{1}{2LfR_L}}v_C(t) \tag{7.32}$$

Analyzing the circuit the transfer function for the converter is given by

$$H(s) = \frac{\tilde{v}_L(s)}{\tilde{v}_C(s)} = \frac{V_S}{V_{st}}\sqrt{\frac{R_L}{2Lf}}\left(\frac{1 + s/\omega_z}{1 + s/\omega_p}\right) \tag{7.33}$$

where $\omega_p$ and $\omega_z$ are given by

$$\omega_p = \frac{2}{R_L C}$$

$$\omega_z = \frac{1}{R_C C} \tag{7.34}$$

and it has been assumed that the equivalent series resistance of the capacitor is very much less than the load resistance. Again, this relationship may be justified using more rigorous methods.

## Summary

This chapter has considered the three basic circuit topologies for switching converters: the buck converter, boost converter and buck–boost converter. In each case the ratio of load voltage to source voltage has been found for both continuous and discontinuous inductor current. It has been shown that for each converter if the current in the inductor is continuous the transformation ratio depends only on the duty ratio, while if the current is discontinuous the ratio depends on both the duty ratio and the load current.

The use of a transformer to isolate the source and load sides of forward and flyback converters has been described briefly. For the forward converter the transformer operates in the usual way, as a device for transforming the voltage and currents while also providing isolation. In the flyback converter the transformer acts as an energy storage device, and is perhaps better considered as two linked inductors.

A simplistic approach has been adopted to obtain the transfer functions from control voltage to load voltage for the buck converter with continuous current and the flyback converter with discontinuous current. These transfer functions will be required in the next chapter to analyze the stability of converters regulated using feedback.

## Self-assessment questions

7.1   A buck converter has a source voltage of 20 V and an inductance of 100 $\mu$H. The switching frequency is 100 kHz. What will be the load voltage if the duty ratio is 0.7, neglecting switch losses, assuming continuous conduction? What is the minimum mean load current for which the inductor current will be continuous?

7.2   A buck regulator is required to produce a load voltage of 10 V from a source voltage in the range 40 to 60 V. If the minimum load current is 250 mA, what is the minimum value for the inductance if the switching frequency is 50 kHz and the inductor current is always to be continuous?

7.3   A boost converter with a switching frequency of 25 kHz steps up a source voltage of 5 V to a load voltage of 15 V. If the inductor has a value of 1 mH, what is the minimum load current for continuous conduction, assuming the converter losses may be neglected?

7.4   A buck–boost switching converter which operates with discontinuous current in the inductor is used to generate a $-15$ V power supply from a $+15$ V supply. If the inductor has a value of 1 mH and the switching frequency is 50 kHz, what will be the required duty ratio if the load current is 10 mA (neglecting losses)?

7.5   For the converter described in Question 7.4, what will be the maximum load current for the inductor current to remain discontinuous? (*Hint:* find the required duty ratio if the current is continuous.)

7.6   A buck–boost converter which operates in discontinuous current is required to supply a load of $-25$ V at 1 A from a source of 15 V. The switching frequency is 100 kHz. How much energy must be stored in the inductor during each cycle? (Neglect losses.)

7.7   The converter described in Question 7.6 has a duty ratio which must be less than the value that would be required if the inductor current were to be continuous. Calculate the maximum time for which the switch may be on, and hence or otherwise deduce the maximum value for the inductor if it is to charge to the required energy.

7.8   An isolated forward converter, as shown in Figure 7.7, has a transformer with 50 turns on the primary winding, 20 turns on the secondary winding and 50 turns on the core reset winding. The source voltage is 100 V. What is the maximum duty ratio? Hence deduce the maximum load voltage (assuming the current in the filter inductor is continuous) and the maximum voltage across the switching transistor.

## Tutorial questions

**7.1** For a buck converter sketch the current waveforms in the inductor, switch and freewheeling diode, assuming the inductor current is continuous.

**7.2** Sketch the switch current and voltage and the diode current for a boost converter in discontinuous current.

**7.3** Explain why, for a boost converter, the mean source current must exceed half the peak-to-peak ripple current in the inductor if the inductor current is to be continuous.

**7.4** Explain why in an isolated forward converter, with a single switching transistor, it is necessary to provide a circuit to reset the flux in the transformer core.

**7.5** For an isolated flyback converter in discontinuous current operation, sketch the currents in the primary and secondary of the transformer, and sketch the voltage across the switching transistor.

**7.6** The transfer function of a buck converter in continuous conduction has a single zero and a pair of complex poles. If the same converter were operated in discontinuous mode, how many poles might be expected? (*Hint:* consider the example of the discontinuous current mode in the flyback converter.)

## References

Chrysis, G., *High Frequency Switching Power Supplies*, McGraw-Hill, New York, 1984.

Kassakian, J. G., Schlecht, M. F. and Verghese, G. C., *Principles of Power Electronics*, Addison-Wesley, Reading, Massachusetts, 1991.

Kislovski, A. S., Redl, R. and Sokal, N. O., *Dynamic Analysis of Switching-Mode DC/DC Converters*, Van Nostrand Reinhold, New York, 1991.

Mohan, N., Undeland, T. M. and Robbins, W. P., *Power Electronics: Converters, Applications and Design*, John Wiley, New York, 1995.

Severns, R. P. and Bloom, E., *Modern DC-to-DC Switchmode Power Converter Circuits*, Van Nostrand Reinhold, New York, 1985.

CHAPTER 8

# Switch-mode power supplies

## 8.1 Introduction

The objective of this chapter is to show how the ideas developed in Chapters 6 and 7 may be put into practice. This is done by presenting two case studies. The objective is to illustrate some of the many issues that must be considered in the design of a switching power supply. By using these case studies it is possible to introduce examples of how concepts introduced in earlier chapters may be used, e.g. driving the power switches, the design of magnetic components and the use of snubbers. A key issue in many power systems is the design of the control system, which usually involves the use of feedback. The design of feedback control systems is complex, and some knowledge of basic control theory is assumed in this chapter. It is also assumed that the reader is familiar with the use of poles and zeros to help describe and analyze control system response. The design technique employed is to use Bode diagrams to help analyze loop performance with a minimum of mathematics. If desired, the sections on the design of the feedback loop may be omitted.

Two examples of switching regulators are considered, and circuits developed that, while not necessarily complete well-protected power supplies, are functional. The first example chosen is a simple buck regulator to take d.c. at between 15 V and 30 V and produce a regulated supply at 5 V, and at a current of between 0.25 and 2 A. The second example is an off-line regulator to operate at an a.c. line voltage of 115 V r.m.s. This regulator uses an isolated flyback topology and delivers a power in excess of 35 W at around 12 V.

Both regulators are designed around a rather old but popular PWM controller IC, the SG3524. This IC is produced by several manufacturers and contains the basic requirements of an oscillator, pulse-width modulator, error amplifiers and voltage reference. More recent control ICs provide more sophisticated facilities, such as soft start, pulse-by-pulse current limiting, under-voltage shut-down and other functions which facilitate the design of reliable and robust switching power supplies. However, the 3524 was chosen as it is simple and provides the basic requirements.

## 8.2  A simple buck regulator

The specification is for a switching regulator to supply 5 V at between 0.25 A and 2

A from a d.c. source at between 15 and 30 V. The initial decision is to choose the topology of the converter. Since the regulator is only required to reduce the voltage a buck (or forward) topology is the obvious choice. Also it is generally best to operate the buck regulator in continuous current mode, as this minimizes ripple on the output voltage and gives good conversion efficiency. There are a number of integrated circuits that provide a simple single-chip solution, some with the power switch included on the chip. The IC chosen here, the SG3524, is a simple low-cost IC that provides all the facilities required, except a power switch capable of supplying sufficient current.

Having selected the topology, the next decision is the operating frequency. This should be chosen so that it is high enough that it will not lead to audible noise (above about 25 kHz). The higher the frequency, the lower the value needed for the inductor, which helps to keep the size to a minimum. However, switching losses will increase with frequency, as in general will the losses in the inductor. A convenient frequency is around 100 kHz, since at this frequency switching losses in the transistor switch or in the freewheeling diode should not be excessive. It is also well within the capability of the 3524 (maximum frequency 300 kHz).

### 8.2.1 Selection of the inductor and capacitor

Assuming that the inductor current is continuous the duty ratio $D$ is given by

$$D = \frac{V_{IN}}{V_{OUT}} \tag{8.1}$$

where $V_{IN}$ is the source voltage and $V_{OUT}$ is the load voltage, 5 V. No allowance has been made for the voltage drop across the switch and the freewheeling diode. The duty ratio may be expected to lie in the range 0.17 to 0.33 depending on the input voltage. If the switching frequency is $f$, the time for which the switch is off each cycle, $t_{off}$, is given by

$$t_{off} = (1 - D)/f \tag{8.2}$$

Neglecting the voltage drop across the freewheeling diode, the current fall in the inductor, $\Delta I_L$, during the off-time is given by

$$\Delta I_L = \frac{V_{OUT} t_{off}}{L} \tag{8.3}$$

The peak-to-peak ripple current must not exceed twice the mean load current if the current in the inductor is to be continuous, so that the minimum value of the inductance is given by

$$L_{min} = \frac{V_{OUT} t_{off}}{2 I_{min}} \tag{8.4}$$

where $I_{min}$ is the minimum value of the mean load current.

Taking the switching frequency as 100 kHz and the minimum value of $D$ gives the maximum off-time as 8.3 $\mu$s. Since the minimum mean load current is 0.25 A, equation (8.4) gives a minimum value for the inductance of 83 $\mu$H. The nearest value of inductance greater than 83 $\mu$H which was readily available as a standard component was 100 $\mu$H, so this value has been chosen. The inductor was rated for 5.4 A d.c., well in excess of the current rating required.

The maximum ripple current is about 0.5 A peak-to-peak, so to keep the ripple voltage down to about 50 mV the impedance of the capacitor at 100 kHz must not exceed 100 m$\Omega$. The value of the capacitance is determined largely by the need to have a low equivalent series resistance (ESR), since at a frequency of 100 kHz the ESR will dominate the impedance for an electrolytic capacitor. Reference to data sheets suggests that a 470 $\mu$F capacitor designed for high-frequency use has an ESR of about 100 m$\Omega$, so this value was chosen. Note that the reactance of a 470 $\mu$F capacitor at 100 kHz is only about 3.4 m$\Omega$.

### 8.2.2 The power switch

At a switching frequency of 100 kHz either a bipolar transistor or a MOSFET can be used as the power switch. Because the regulator was to be designed with a common negative line (or ground) a $p$-channel MOSFET would be required in order to avoid the need for isolation of the gate drive circuit, with its added complications. As $p$-channel MOSFETs tend to be more expensive than $n$-channel devices, it was decided to use a bipolar device instead. Rather than use a $p$–$n$–$p$ device as the power switch, a complementary feedback pair was used as shown in Figure 8.1.

The $n$–$p$–$n$ power transistor, Q1, cannot saturate since its collector potential cannot fall below the base potential. However, the $p$–$n$–$p$ drive transistor, Q2, could be driven into saturation. In order to avoid this, and the consequent storage delay, a Baker's clamp comprising diodes D1 and D2 is used to clamp the collector voltage to the base voltage. The base of Q2 is current driven through the 2.2 k$\Omega$ resistor so

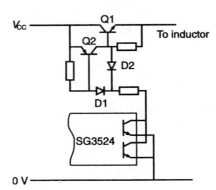

**Figure 8.1** *The bipolar transistor switch drive circuit for the buck regulator*

that changes in supply voltage may be accommodated. This base drive circuit is simple, but is not ideal, since it cannot provide a reverse bias to the bases of the transistors at turn-off. The two output transistors of the 3524 are active on alternate cycles (to facilitate the implementation of push–pull switching circuits). By connecting them in parallel, a pulse is obtained for every cycle of the oscillator.

The freewheeling diode must be either a Schottky diode or a fast-recovery junction diode with a reverse voltage of at least 30 V. A BYW98 fast-recovery diode with a forward current of 3 A and a reverse voltage of 150 V was chosen.

### 8.2.3 Closing the feedback loop

In order to design the compensation for the feedback loop it is necessary to model all the circuit within the feedback loop. The error amplifier is a transconductance amplifier with a transconductance of 2 mS and an effective load resistance of 5 MΩ giving an effective voltage gain at low frequency of 80 dB. The gain of the amplifier has fallen by about 3 dB at 300 Hz and the roll-off shows a single-pole response. A simple linear model of the amplifier is a voltage-controlled current source with a transconductance of 2 mS in parallel with a load resistance of 5 MΩ and a capacitance of 106 pF. Since the common-mode voltage of the amplifier must be limited to the range 1.8 to 3.4 V, it is necessary to divide the internal 5 V reference voltage of the IC by 2. The required load voltage is also 5 V, hence, the output voltage must also be divided by 2.

The transfer function of the power stage may be obtained from equation (7.29) of Chapter 7. The frequency of the pair of complex poles, $\omega_0$, is 461 rad s$^{-1}$, or 734 Hz and the value of $Q$ is 4.6, assuming that the equivalent series resistance (ESR) of the capacitor is 100 mΩ. The zero occurs at a frequency of 3.39 kHz, assuming a capacitance of 470 $\mu$F and an ESR of 100 mΩ. The effective gain of the power stage depends on the ratio of the source voltage to the amplitude of the sawtooth wave form of the pulse-width modulator, $V_{st}$. The data sheet gives the value of $V_{st}$ as 3.5 V, hence the gain will lie in the range of 4.29 at a supply of 15 V to 8.57 at 30 V.

The two complex poles will introduce a phase shift of $-180°$ as the frequency is increased from a low value through $\omega_0$ to a high value, although the phase will be pulled back from $-180°$ by the zero due to the ESR of the capacitor. A Bode diagram of the power stage transfer characteristic is shown in Figure 8.2, assuming a source voltage of 20 V. It is clear that the large phase shift associated with this pair of complex poles may introduce instability when the feedback loop is closed, unless care is taken.

The simplest strategy for compensating the feedback amplifier is to introduce a low-frequency pole and a zero at a frequency somewhat below the resonant frequency. This is easily achieved with the transconductance amplifier of the 3524 by connecting a resistance and a capacitance in series across the compensation pin (the amplifier output) and the ground. The frequency of the zero was chosen to be at about half the resonant frequency of the filter. The value of the resistance should be not less than 30 kΩ to avoid limiting the voltage swing of the amplifier. A value

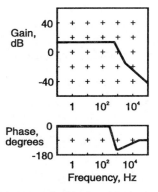

**Figure 8.2** *Bode plot for the transfer function of the buck converter power stage*

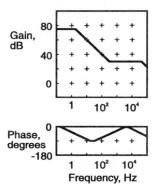

**Figure 8.3** *Bode plot of the feedback amplifier transfer function*

of 33 kΩ was chosen with a capacitance of 15 nF to give a zero frequency of 321 Hz, and a pole at 2.2 Hz. The Bode plot of the transfer function of the error amplifier is shown in Figure 8.3.

Combining the Bode plots in Figures 8.2 and 8.3 gives a Bode plot for the loop gain of the feedback for the regulator (Figure 8.4). It can be seen that while the phase shift reaches almost 180° just above the resonant frequency of the filter, the phase is pulled back to allow a comfortable phase margin at the unity gain frequency (about 20 kHz). It is interesting to note that this phase margin is only obtained because of the zero produced by the ESR of the filter capacitor. Without this zero, it would be necessary to add another zero to the error amplifier compensation.

To confirm the performance of the feedback loop the amplifier and power stage have been modelled using the circuit shown in Figure 8.5. The SPICE listing is given in Appendix 1 (Section A1.5). The power stage gain is for a supply voltage of 20 V. This was analyzed using SPICE and the calculated transfer characteristics for the loop are shown in Figure 8.6. The predicted phase margin is about 60°, which

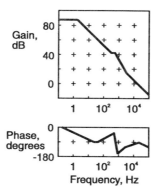

**Figure 8.4** *Bode plot of the trnasfer function of the buck regulator*

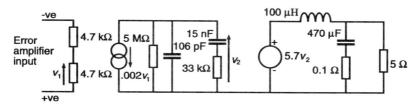

**Figure 8.5** *Circuit model for the error amplifier and power stage of the buck regulator*

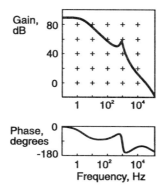

**Figure 8.6** *SPICE simulation of the loop gain of the buck regulator*

should provide a reasonable step response. Reducing the ESR of the capacitor, from 100 mΩ to 10 mΩ, gave a phase margin of only 3°, which would give a feedback loop on the verge of instability. A safer design strategy would be to introduce another zero in the compensation. In practice, this does not seem to be necessary.

Having set up a SPICE simulation of the circuit model shown in Figure 8.5, it is easy to close the loop and test the response to a step change in the load current. This was done by connecting the inverting input node to the top of the 5 Ω load resistor

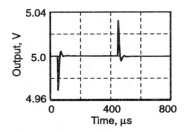

**Figure 8.7** *Simulated step response of the closed-loop converter*

and the non-inverting node to a 2.5 V source. The step in load current was simulated by switching an extra load resistance of 15 $\Omega$ in parallel with the 5 $\Omega$ load. The results are shown in Figure 8.7. This shows that a reasonable step response may be expected.

### 8.2.4 Testing the design

The final circuit for the regulator is shown in Figure 8.8. The circuit was constructed, using bread-board techniques, and tested. The regulator produced 5 V as expected, and operated satisfactorily with a load current of up to 2 A, and with the source voltage in the range 15–30 V. The efficiency with a source voltage of 20 V and a load current of 1 A was measured to be 70%, decreasing to 67% with a source voltage of 30 V. The efficiency also decreased slightly with increased load current.

In order to investigate the sources of power loss it is necessary to measure the current in the switches. The high-frequency current waveform was measured using

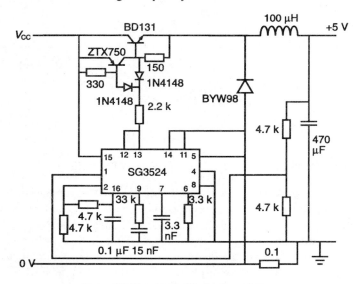

**Figure 8.8** *The circuit of the buck regulator*

a current transformer. This was constructed using a ferrite toroid with a single-turn primary, consisting of the wire carrying the current to be measured, and a 25-turn secondary. The toroid had a magnetic path length of 57 mm, an effective cross-sectional area of 31.5 mm and was made from 3E2 grade ferrite. The secondary inductance was about 2 mH and was connected to a resistive burden of 10 $\Omega$, giving a primary impedance of only around 20 m$\Omega$. A resistance of 50 $\Omega$ was used in series with the coaxial cable to the oscilloscope to damp ringing in the cable.

The current in the freewheeling diode was measured by passing the diode through the toroid. The voltage across the diode and the current through it are shown in Figure 8.9. These waveforms were obtained with a source voltage of 20 V and a load voltage of 5 V. The sampling rate of the digital oscilloscope was not fast enough to record the detail of the rising and falling edges of the current and voltage on this time-base range. The measured switch duty ratio is about 0.33, much greater than the ratio of the load-to-source voltages (0.25). The decrease in inductor current during the freewheeling phase is 0.36 A in 6.3 $\mu$s, or $5.7 \times 10^4$ As$^{-1}$, which is consistent with the load voltage of 5 V, the diode forward voltage of about 0.7 V and the inductance of 100 $\mu$H.

That the duty ratio exceeds the predicted value is due to the power losses in power switch and diode. It is useful to consider where this power is lost. The diode has a forward voltage of 0.75 V and a mean load current of 1 A for two-thirds of the cycle, corresponding to a mean on-state loss of 0.5 W. The power switch has a forward voltage of about 1.4 V and carries a mean current of 1 A for one-third of the cycle, corresponding to a loss of 0.467 W. The drive for the power switch from the PWM controller has a current of about 9 mA, again for one-third of the cycle, corresponding to a mean power of about 60 mW, while the IC itself draws about 180 mW. Thus the total power loss from static and on-state losses is about 1.21 W. The actual power loss deduced from the difference between the input power and the power delivered to the load is 2.1 W.

In order to estimate the switching losses, the voltage across the freewheeling diode and the current through it have been recorded, as shown in Figure 8.10 for the switch turn-on transition. The current in the diode just before the transition is 0.82

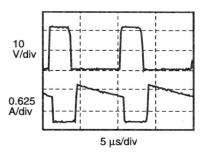

**Figure 8.9** *Upper trace, the voltage across the freewheeling diode (10 V/div); lower trace, the current through the freewheeling diode (0.625 A/div); timescale 5 μs/div*

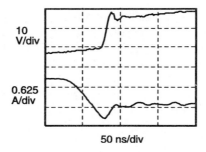

**Figure 8.10** *Upper trace, the voltage across the freewheeling diode at switch turn-on (10 v/div); lower trace, the current through the freewheeling diode (0.625 A/div); timescale 50 ns/div*

A, and the current reversal in the diode during reverse recovery can clearly be seen. The current in the switch rises from zero to 1.25 A, before the diode turns off and the voltage collapses. This transition lasts about 60 ns. The energy, $E$, lost during the transition may be estimated from equation (6.6) (in Chapter 6), which gives

$$E = 0.5t_r V_{CC} I_{max}$$

Substituting for $t_r$, $V_{CC}$ and $I_{max}$ gives an energy loss of about 750 nJ. At a frequency of 100 kHz the turn-on loss will contribute about 75 mW to the total losses.

The voltage across the freewheeling diode and the current through it as the switch turns off are shown in Figure 8.11. The reverse voltage across the diode does not fall rapidly as might be expected from the discussion in Chapter 6. The current in the switching transistor falls very slowly at first, with a consequence that the diode reverse voltage falls slowly for about 0.5 $\mu$s, as the stray capacitance, and the capacitance of the diode is discharged. While the voltage falls the current in the transistor will remain almost constant at about 1.2 A. The area under the switch voltage as it increases may be estimated by graphical integration, and the energy lost found by multiplying the area by the current. This gives an energy of 3 $\mu$J. The rise time of the current in the diode is about 100 ns, hence the energy lost during

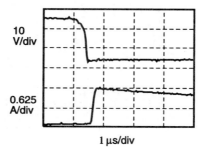

**Figure 8.11** *Upper trace, the voltage across the freewheeling diode at switch turn-off (10 V/div); lower trace, the current through the freewheeling diode (0.625 A/div); timescale 1 $\mu$s/div*

this phase can be calculated in the same way as for turn-on, and is about 1.2 $\mu$J. Thus the total energy lost at turn-off is 4.2 $\mu$J, and the power lost is 0.42 W.

The total losses accounted for are about 0.5 W for switching losses, 0.97 W for on-state losses and 0.24 W for power dissipated in the IC and the switch driver. This leaves about 0.39 W of loss unaccounted for. The most likely explanation of this loss is that it is due to the hysteresis and copper losses in the inductor and losses in the capacitor.

In order to test the response of the regulator to step changes in load current a 15 $\Omega$ load resistor was switched in parallel with the 5 $\Omega$ load resistance to produce a step change of 0.33 A. The response of the load voltage is shown in Figure 8.12. The ripple voltage due to the switching has a peak-to-peak amplitude of about 20 mV, giving an ESR of about 55 m$\Omega$ for the capacitor, rather than the 100 m$\Omega$ assumed. The peak value of the transient is much higher than predicted by the simple theory (Figure 8.7) and it is also variable depending on when the transient occurs with respect to the switching cycle. It is clear that the response time is such that it is not possible to assume that voltages and currents change slowly on the timescale of the switching period, so better agreement should not be expected. The finite output impedance, about 60 m$\Omega$, arises from the losses in the converter which require a change in duty ratio with a change in steady-state load (not required for an ideal buck regulator with continuous current).

Overload protection is provided by an amplifier which senses the voltage across the 0.1 $\Omega$ resistor in series with the negative supply. If this exceeds 200 mV the pulse width is reduced, limiting the current. This current-limiting scheme has two disadvantages: it must have a resistance in series with the negative supply, which ceases to be a common line, and it does not limit the peak switch current, only the mean load current.

### 8.2.5 Review of the buck regulator design

The design that has been considered is for a basic buck regulator delivering up to 10 W at 5 V. With a 20 V source voltage the current in the inductor remains

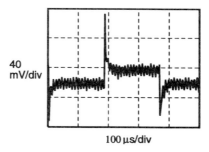

40 mV/div

100 µs/div

**Figure 8.12** *The response of the buck regulator to a step change in load; vertical scale 40 mV/div; timescale 100 µs/div*

continuous down to a load current of 0.2 A, although it will work quite satisfactorily with a much smaller load current. The efficiency has a maximum value of about 70%. Analysis shows that the losses are reasonably evenly distributed between several sources. Small improvements in efficiency are possible by using a Schottky diode to reduce the diode forward voltage, and by using a faster switch, perhaps a MOSFET, to reduce switching times and possibly the on-state voltage.

The stability of the feedback loop has been considered in some detail because of the complexity introduced by the poles of the LC filter. The simple pole and zero compensation used in this example is usually adequate to obtain stability of the loop for the buck regulator. One disadvantage of the control scheme described is that while it gives good regulation to changes in load current, it gives poorer regulation with respect to changes in the source voltage. This is because for changes in the load current the duty ratio will only need to change slightly to allow for changes in the power loss, while changes in source voltage require larger changes in duty ratio. What is needed is *feedforward* to correct immediately for changes in source voltage.

There are alternative control strategies which may provide better closed-loop performance, for example the use of *current mode* control of the pulse width. In this mode of operation the pulse width is not controlled directly by the control voltage, instead the pulse-width controller controls the maximum inductor current. This strategy has two advantages: it removes the complex poles from the feedback loop, and replaces them with a single real pole, and it provides feedforward.

## 8.3  An isolated flyback converter

The second design study is for an isolated converter to deliver a power of at least 35 W at about 12 V. The power source is to be a.c. at a nominal voltage of 115 V r.m.s. This is an example of an off-line switching power supply, in which the input to the rectifier may be connected directly to the a.c. supply. The nominal supply voltage of 115 V was chosen for reasons of safety when testing the experimental design and also to enable a low-cost MOSFET switch to be used. A 220–240 V supply would require a switch with a voltage rating of at least 700 V.

The simplest topology for low-power off-line switching power supplies is the isolated flyback circuit. It uses the fewest components of any of the isolated switching converters and, if used in discontinuous conduction, the feedback is easy to implement. For these reasons this topology was chosen.

The off-line switching regulator must obtain power for the PWM controller and the switch drive, at a voltage substantially less than that of of the rectified d.c. A simple way to achieve this is to include a tertiary winding on the transformer (or coupled inductor), which will supply power at about 12–15 V for the control circuit. The basic circuit for the power supply is shown in Figure 8.13, which is based on the circuit described in Chapter 7, Section 7.5.3.

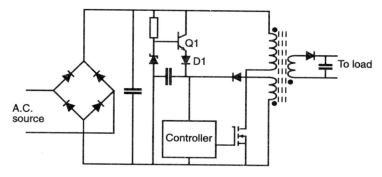

**Figure 8.13** *Basic circuit of the off-line, flyback regulator*

At start-up the auxiliary power will not be available, so a 'kick-start' circuit is required to supply power until the switching supply is operating. This is achieved by the transistor Q1 which supplies power at about 11.5 V to the controller until the power supplied from the switching supply reverse biases D1, cutting off the direct supply.

### 8.3.1 Design of the energy storage transformer

The key component in the isolated flyback converter is the energy storage transformer, or coupled inductor. It is, therefore, appropriate to consider this component in some detail. In order to avoid a rather complex transfer function, which is difficult to compensate so as to obtain stable feedback, it is usual to operate the transformer with discontinuous current. This means that the secondary current(s) must have fallen to zero before the primary winding is energized at the start of the next cycle. Alternatively, the energy storage transformer stores energy while the primary is energized, then transfers all this energy to the load during the second part of the cycle.

The energy stored per cycle, $E$, multiplied by the switching frequency, $f$, gives the maximum possible power transferred to the load. In practice the converter may have an efficiency of only about 70%, so the power transfer may only be about $0.7Ef$. If the required power transfer is 35 W and the frequency 50 kHz, then, allowing for an efficiency of only 70%, the required energy that must be stored each cycle is 1 mJ. The stored energy is one of the factors that determines the size of core required, and the others are an adequate window area for the windings and adequate heat dissipation to limit the temperature rise.

The design of an energy storage transformer starts with the choice of core. The optimum choice of ferrite, core size and geometry is complex and beyond the scope of this book. Here an ETD39 core made from ferroxcube 3C8 ferrite was chosen. This core is rather larger than required, but was readily available. The ferrite is suitable for use at frequencies between about 20 and 150 kHz. The core is constructed from two E-sections, which may be separated by a suitable spacer to provide an air gap.

The data sheet for the core specifies the essential geometric and magnetic parameters of the core:

Magnetic path length, $l = 92.2$ mm
Effective area of magnetic path, $A = 125$ mm$^2$
Effective magnetic volume $= 11\,500$ mm$^3$
Specific inductance $= 2.7\ \mu$H per turn$^2$
Saturation flux at 100° C $= 0.32T$.

The specific inductance is the reciprocal of the reluctance which is given by

$$\Re = \frac{l}{\mu\mu_0 A}$$

Using the values specified for the core gives a value of 0.002 for $\mu\mu_0$. This allows the maximum energy that may be stored in the core, $E_{cm}$, to be calculated from

$$E_{cm} = \int_V 0.5 HB dV = \frac{V B_{max}^2}{2\mu\mu_0}$$

where $V$ is the core volume and $B_{max}$ is the maximum working flux. Taking the maximum working flux as $0.3T$ gives the maximum energy stored in the core as 260 $\mu$J. Any extra energy must be stored in the air gap.

The maximum energy stored in the air gap, $E_{gm}$, is calculated from

$$E_{gm} = \frac{V_g B_{max}^2}{2\mu_0}$$

where $V_g$ is the effective volume of the gap, the effective area multiplied by twice the gap width $g$. The factor of two arises because there are two gaps in series, one in the pole and one in each side-limb of the core. The pole area and the total area of the side-limbs are both approximately equal to the effective core area. Substituting for the maximum working flux density, for the core area and for $\mu_0$ gives the maximum stored energy in the gap as $8.95g$. If the maximum total stored energy is 1 mJ, then the maximum energy stored in the gap is 0.74 mJ and the gap width 83 $\mu$m. The spacer used was a piece of mylar film, and the specific inductance of the core with the spacer in place was measured to be 637 nH per turn. The thickness of the spacer deduced from the specific inductance was 94 $\mu$m, giving the maximum stored energy as 1.1 mJ.

The primary of the transformer was wound with 21 turns of two strands of 0.56 mm copper wire connected in parallel, which gave a primary inductance of 280 $\mu$H. This will give an energy storage of 1 mJ at a current of 2.67 A. Assuming that the d.c. input to the regulator is at a voltage between 130 and 160 V the time taken to charge the inductor to this value is between 4.7 and 5.8 $\mu$s. This leaves about 14 $\mu$s to discharge the inductor while the switch is off.

The secondary was wound with four turns to give an inductance of 10.1 $\mu$H. When the primary switch is turned off the current flows in the secondary. Since there are fewer turns the secondary current will be 21/4 times greater than the primary current, neglecting any leakage of flux. If the final primary current is 2.67 A, the initial secondary current must be 14 A, to support the flux. With a secondary voltage of 12 V the current will decay in about 12 $\mu$s, confirming that a four-turn secondary is satisfactory.

In practice, the tertiary winding, which had the same number of turns as the secondary, was wound over the primary, insulated by several layers of tape. The wire diameter was 0.56 mm. The primary and tertiary windings were insulated with several layers of mylar film, then the secondary winding was wound using a conductor formed from seven strands of insulated 0.56 mm wire twisted together. This *Litz wire* construction helps to minimize the resistance at the switching frequency, where the skin depth is small.

The transformer is not of optimum size, since the windings do not fill the window of the core, and in use the temperature rise of the core is small. A smaller size of core could have been used. In a transformer for use in an off-line application it is necessary to pay particular attention to the insulation. The thickness of insulation and the creepage paths must meet the appropriate requirements to ensure adequate isolation of the primary and tertiary windings from the secondary winding.

### 8.3.2 Leakage inductance and snubbers

Not all the flux in the primary winding of the energy storage inductor will be linked with the secondary and tertiary windings, leading to leakage inductance. When the switch in the primary is opened a path must be provided to discharge the energy stored in the primary leakage inductance. This may be achieved using D1, C1 and R1 in the circuit shown in Figure 8.14.

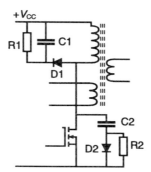

**Figure 8.14** *Snubber and clamp circuits for the flyback regulator*

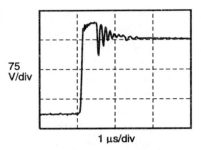

75
V/div

1 μs/div

**Figure 8.15** *The voltage across the MOSFET as it turns off; vertical scale 75 V/div, timescale 1 μs/div*

When the switch turns off the voltage across the primary winding reverses, and with no leakage inductance would have a value of $nV_s$, where $n$ is the ratio of primary turns to secondary turns for the transformer and $V_s$ is the secondary voltage, which is clamped by the reservoir capacitor and the load. When there is leakage inductance the primary voltage will increase further to maintain the current in the leakage inductance. This charges the capacitor C1 through the diode D1, absorbing the energy stored in the leakage inductor and clamping the peak switch voltage. When current in the leakage inductance falls to zero the voltage across the primary winding falls back to $nV_s$. The energy stored in the capacitor is dissipated in R1. The value of the capacitor and the time constant are chosen so that the capacitor voltage remains above $nV_s$ throughout the cycle. This ensures that the diode D1 only conducts to discharge energy transferred to the capacitor from the leakage inductance.

The values chosen for C1 and R1 were 33 nF and 4.7 kΩ. The measured rising edge of the voltage waveform across the switch is shown in Figure 8.15, with the supply connected to a load resistance of 4.7 Ω. A larger value of R1 would probably be desirable to discharge the leakage inductance more rapidly, at the expense of a higher peak voltage across the switch.

The turn-off snubber, which comprises C2, R2 and D2 in Figure 8.14, limits the rate of rise of voltage across the switching transistor. It also serves to damp the ringing when the clamp diode D1, or the main diode in the secondary circuit, ceases conduction. When D1 cuts off, the voltage across the primary exceeds the value determined by the e.m.f. induced by the current flowing in the secondary. As the voltage tries to relax to its equilibrium value, the energy stored in the stray capacitance will resonate with the leakage inductance, and produce ringing. The snubber adds extra capacitance, C2 , but it also adds damping due to R2. The ringing observed when the clamp diode cuts off is clearly seen in Figure 8.15. A similar effect occurs when the secondary current ceases. In this case the total primary inductance resonates with C2 and the stray capacitance of both the primary and secondary circuits. Adequate damping of these resonances helps to minimize electromagnetic interference, and can also help to avoid problems with the control loop. The values chosen must provide a reasonable compromise of adequate damping while avoiding excessive energy loss in the snubber.

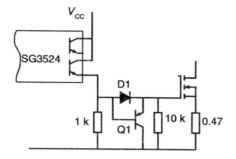

**Figure 8.16**  *The gate drive circuit for the MOSFET switch*

### 8.3.3 The power switch

The maximum d.c. voltage after the rectifier is about 170 V. When the current is flowing in the secondary there will be a reverse voltage of around 80 V across the primary, giving a maximum voltage across the switch of 250 V. This may be substantially greater while the energy stored in the primary leakage inductance is discharged, so a reasonable minimum voltage rating for the switch is 400 V. The device chosen was an IRF830, a low-cost MOSFET with a maximum drain voltage of 500 V and a maximum drain current of 4.5 A.

The output stage of SG3524 provides a transistor which may be used as a common collector stage to drive the MOSFET. This will provide a fast turn-on. To obtain a fast turn-off the circuit shown in Figure 8.16 was used. At turn-on, the MOSFET gate is pulled up through the diode and the transistor Q1 is cut off. At turn off the transistor Q1 turns on rapidly discharging the gate. Note that only one of the output transistors of the 3524 is used. This limits the duty ratio to less than 50%, and requires that the oscillator frequency be twice the desired switching frequency.

### 8.3.4 Control

The control loop needs isolation between the control IC and the load circuit. There are a variety of ways of achieving this. Two popular methods are to use optical isolation or a modulated high-frequency carrier and an isolation transformer. Here a much simpler if less effective method was chosen. The feedback loop controls the voltage of the auxiliary supply to the pulse-width modulator. If there were no leakage inductance and no resistive losses in the secondary windings this would give good control of the load voltage. To improve the tracking of the voltage across the two secondary windings, a resistance with a value of 100 $\Omega$ is connected across the auxiliary supply. This minimizes the effect of the leakage inductance spike.

As discussed in Chapter 7, the flyback converter power stage has a simple transfer characteristic, with a single pole at a frequency $(R_L C_L)^{-1}$, where $R_L$ is the

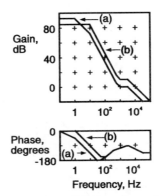

**Figure 8.17**  *A Bode plot of the loop gain of the flyback regulator with a load of (a) 35 Ω and (b) 3.5 Ω*

load resistance and $C_L$ is the reservoir capacitance. The load resistance and capacitance on the auxiliary winding can be assumed to be in parallel with the load resistance and capacitance for the purpose of simple analysis. The gain of the converter is given by equation (7.33), neglecting the transformer. If the transformer is included the voltage $V_S$ must be scaled by the transformer ratio (4/21), and the inductance will be the secondary inductance. Since only one of the output transistors of the control IC is used, the power switch is only operated on alternate cycles of the pulse-width modulator. This limits the maximum duty ratio, and the the variation of duty ratio with control voltage will be reduced by a factor of two, or the value of $V_{st}$ used in the calculations must be replaced by $2V_{st}$. Assuming that the load lies in the range 3.5 to 35 Ω, the power stage gain is in the range of 24 at 35 Ω down to 7.5 at 3.5 Ω. The reservoir capacitor used has a value of 4700 $\mu$F, giving a pole at 0.97 Hz for the 35 Ω load, and 9.7 Hz for the 3.5 Ω load. The ESR of the capacitor will lead to a zero in the transfer function at a frequency of around 1 kHz.

The amplifier is compensated with a pole at 10 Hz and a zero at 500 Hz using a 100 kΩ resistance in series with a 3.3 nF capacitor between the compensation pin and ground. To ensure that the loop gain has fallen to zero at a frequency well below the switching frequency a further pole is introduced at a frequency of 3.4 kHz by a further capacitor of 470 pF between the compensation pin and ground. A Bode plot of the loop gain for the two values of load resistance is shown in Figure 8.17.

### 8.3.5  Testing the final circuit

The final circuit is shown in Figure 8.18. The supply was tested with a nominal 115 V at 50 Hz a.c. source and load values between 60 Ω and 3.35 Ω. The mean load voltage decreased from about 14.4 V with the 60 Ω load to 11.6 V at a load of 3.4 Ω. The output resistance was almost constant at 0.9 Ω across this range of load. The auxiliary voltage supply remained almost constant as the load was varied, confirming that the voltage drop is because the voltages across the two secondaries

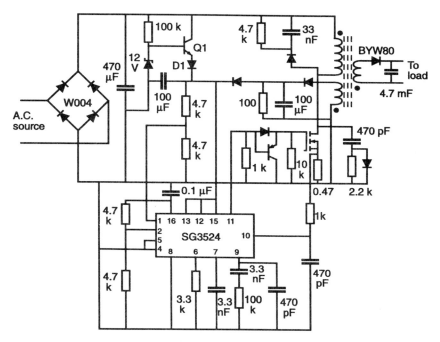

**Figure 8.18** *The final circuit of the flyback regulator*

do not track very well. A better feedback isolation technique is clearly required if voltage stability is important.

The efficiency of the converter was measured as the ratio of power in the load to the d.c. power supplied from the rectifier. The efficiency increases with load, with a maximum value in excess of 75% when the load resistance is 3.5 Ω. Significant power is lost in the clamp resistor which dissipates the energy stored in the leakage inductance (2 or 3 W at full load) and the resistor across the auxiliary supply (about 2 W).

The voltage across the switch and the switch current are shown in Figure 18.19. This is with a load of 4.7 Ω. Notice the step in the current as the switch turns on. This is the turn-off snubber discharging its stored energy. The heavily damped oscillation of the primary inductance with the snubber capacitor can be seen when the load current is extinguished and the switch voltage falls back to the value of the d.c. supply.

## 8.3.6 Review of the flyback regulator design

The flyback regulator described performs adequately for the purpose of illustrating the principles. The circuit is not a complete switching power supply. The current limiting for the MOSFET uses the shut-down pin on the 3524 and is rather

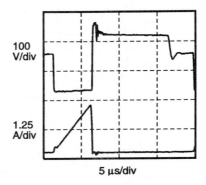

**Figure 8.19** *(a) The voltage across the MOSFET switch (100 V/div); (b) the current through the MOSFET (1.25 A/div); timescale 5 μs/div*

rudimentary. Extra filtering is needed at the output to reduce voltage ripple and an input filter is required to block high-frequency emissions from reaching the main supply.

Efficiency is quite reasonable but load regulation is poor. This could be corrected by using an isolated feedback circuit to control the load voltage directly, rather than control the auxiliary supply. The transformer is certainly less than optimum. The size of core used should permit a power transfer in excess of 100 W. With the design discussed here, the maximum power is limited by the inductor current becoming continuous (which makes the feedback loop unstable).

Designing an off-line, flyback regulator for use with a 240 V supply is no different in principle, but a switch with a significantly higher voltage rating is required. A high-performance bipolar transistor might be preferred to a MOSFET in these circumstances. Alternatively, a different circuit topology, such as a full- or half-bridge forward converter might be preferred, since for these circuits the voltage rating of the switching transistors need not be much greater than the maximum supply voltage.

## 8.4  Control of electromagnetic emissions

Switching power supplies generate large high-frequency circulating currents. Careful design is required to minimize the magnitude of such currents, by ensuring that the recovery time of diodes is sufficiently fast and that switches which can short the supply are never on simultaneously, even for very short times. It will also help if the rate of rise of current and voltage is controlled by the use of snubbers, or using more complex *resonant* designs where the switch is operated at zero current or voltage. However, despite such measures, it is still necessary to contain the unwanted RF emissions.

Good layout with careful attention to where the high-frequency currents must flow is essential. The area of loops in which the current must flow should be minimized together with the stray magnetic fields due to flux leakage from inductors. Earth loops are a particular problem. Any wiring or metallic components that are subjected to voltages that change abruptly as the converter operates should be made as small as possible since even small, stray capacitance may introduce significant circulating currents. Electrostatic shielding will help but careful choice of the earthing point is necessary.

The supply lines to the converter should be filtered, as close as possible to the input of the converter. The circuit of Figure 8.20 is frequently used. The series inductances, which might have a value of about 100 $\mu$H, provide a high impedance for current circulating through the power supply and also for common-mode current circulating through both power leads and returning via ground. The common-mode signal and its associated earth return current tends to be a particular problem with switching power supplies. Sometimes coupled inductors are used to provide a large impedance to common-mode current but a smaller impedance to circulating current. With switching supplies much of the noise is likely to be at a very high frequency (around 10 MHz), and the inductors should therefore be wound on a core material suitable for use at this frequency. If lower-frequency signals are also to be blocked (at, say, the switching frequency of around 100 kHz) then it may be advisable to have two inductors in series, one a high-frequency but low-value inductor and the second a higher-value inductance but with poorer high frequency performance.

The capacitors across the power rails provide a low-impedance path for the high-frequency current circulating through the power rails from the switching supply, while the capacitors to ground pass the circulating common-mode current. The shunt capacitors across the power lines may have a value of around 0.1 $\mu$F while those to ground may have a value of around 10 nF. The location of the earth return connection for the capacitors carrying the common-mode current needs to be chosen with care. The current will return via stray capacitance between those parts of the circuit that have rapidly changing voltage and ground. If a high-frequency transformer with an electrostatic shield between the primary and secondary is used then this shield will carry the bulk of the circulating current. Hence the capacitors should be returned to this shield. Most of the current circulating through the power

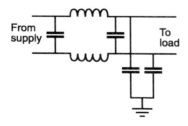

**Figure 8.20** *Filter circuit for power lines*

lines to ground will return via the capacitance between primary winding, with its large, rapidly changing voltage, and the shield.

## Tutorial questions

**8.1**  For a buck converter what factors should be taken into acount when considering the switching frequency?

**8.2**  For a buck regulator, how will on-state losses in the power switch and freewheeling diode affect the duty ratio required to achieve a given output voltage when compared to an ideal converter? Explain your answer.

**8.3**  Consider how the bipolar transistor in Figure 8.1 might be replaced by a *p*-channel power MOSFET. How could a satisfactory switching speed be obtained while limiting the maximum gate–source voltage?

**8.4**  Why is the equivalent series resistance (ESR) of the reservoir capacitor important in ensuring stability of the feedback loop for a buck regulator?

**8.5**  What will be the effect of slow reverse recovery of the freewheeling diode on a buck regulator?

**8.6**  How and why does the energy storage transformer of an isolated flyback regulator differ from a normal high-frequency transformer?

**8.7**  Why is an auxilliary supply derived from the power transformer used to power the control chip in the isolated flyback regulator described?

**8.8**  Explain how the clamp circuit shown in Figure 8.14 (D1, C1 and R1) absorbs the energy stored in the primary leakage inductance of the transformer.

**8.9**  Why is a turn-off snubber used in the flyback regulator (C2, R2 and D2 in Figure 8.14)?

## References

Mohan, N., Undeland, T. M. and Robbins, W. P., *Power Electronics: Converters, Applications and Design*, John Wiley, New York, 1995.

*Unitrode Switching Regulated Power Supply Design Seminar Manual*, Unitrode Integrated Circuits Corporation, Merrimack, 1990.

*Unitrode Linear Integrated Circuits Data and Applications Handbook*, Unitrode Integrated Circuits Corporation, Merrimack, 1990.

## CHAPTER 9

# *Controlled rectification*

## 9.1 Introduction

Thyristors have, over many years, established themselves as robust and reliable power switching devices, which are available in very large voltage ratings (up to about 5 kV) and with current ratings in excess of 1000 A. They are most suitable for use at power-line frequency (50 or 60 Hz) where their relatively slow switching speed is not a problem. In circuits designed to control relatively low power, they have to a large extent been replaced by other devices which are able to switch more rapidly and can be controlled more easily. However, one group of circuits, the controlled rectifiers, are particularly well suited to thyristors. A controlled rectifier is a circuit for converting a.c. to d.c. in which controlled switches, usually thyristors, replace some or all of the diodes of a simple rectifier. This allows the mean d.c. voltage to be controlled by controlling the times at which the switches are turned on.

This chapter aims to explore single-phase controlled rectifier circuits. In practice, controlled rectifiers are now used mostly at a power level of many kilowatts, other methods of control having become prevalent at lower power. For this reason, the three-phase versions of the circuit are probably used more widely than the single-phase circuits. However, the essential characteristics of the controlled rectifier can be studied just as well using the simpler single-phase circuit.

Rectifiers convert a.c. to d.c. but some of the controlled rectifier circuits are also able to produce a power flow in the reverse direction (d.c. to a.c.). This process, the conversion of d.c. to a.c. is referred to as *inversion*. Rectification and inversion are two examples of *power conversion*, which is the generic term used to describe the transformation of electrical power at a particular voltage level and frequency (including d.c.) to another voltage level and/or frequency. To clarify their bi-directional capability, controlled rectifiers are frequently referred to as controlled converters. However, in this chapter the circuits will be considered primarily as rectifiers.

## 9.2 The half-wave controlled rectifier

The simplest possible controlled rectifier is the half-wave circuit shown in Figure 9.1. Controlled rectifiers supplied from an alternating voltage source of low

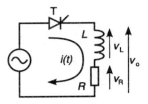

**Figure 9.1** *A simple half-wave controlled rectifier*

impedance must always be used with a load that is either a pure resistance or inductive. It is clearly unwise to switch a voltage across a capacitive load, since a large impulsive current will flow. Here the load is assumed to have an inductance $L$ and a series resistance $R$. The thyristor can only conduct if it is forward biased, and then only when it receives a trigger pulse. The source voltage is assumed to be sinusoidal with a frequency, $\omega$, and an r.m.s. magnitude $V$. The thyristor is triggered during the positive half-cycle, and the trigger pulse occurs after a phase delay of $\alpha$, referred to as the voltage zero.

Neglecting the thyristor forward voltage, the current while the thyristor is conducting is given by

$$L\frac{di}{dt} + iR = \sqrt{2}V \sin \omega t \tag{9.1}$$

where $V$ is the r.m.s. value of the supply voltage.

The solution of this equation is shown graphically in Figure 9.2. The current is shown in the figure as the product of the current, $i$, and the resistive part of the load, $R$. This method of representing the current permits the display of the current and voltage on the same axes. The load voltage, $v_0$, follows the sinusoidal source voltage while the thyristor is conducting ($i > 0$). The difference between the two curves is the voltage across the inductance, $v_L$. If $v_L$ is positive the current will be increasing, while if $v_L$ is negative it will be decreasing. The rate of increase or decrease of the current is proportional to $v_L$. The two shaded areas between the curves must be equal, since the time-averaged voltage across the inductance is zero. Using these two properties makes it possible to sketch the current. Note that for part of the cycle the load voltage is negative while the current is still positive. This implies that energy stored in the inductor is being returned to the supply.

Inspection of Figure 9.2 shows that since the current is always positive it must fall to zero before the end of the cycle in order to satisfy the equal area condition. Hence, the initial condition is that $i = 0$ when $\omega t = \alpha$. and equation (9.1) may then be solved to obtain

$$i(t) = \frac{\sqrt{2}V}{Z}\left(\sin(\omega t - \phi) - \sin(\alpha - \phi)\exp\left(\frac{R(\alpha - \omega t)}{\omega L}\right)\right) \tag{9.2}$$

where

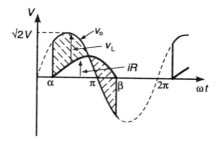

**Figure 9.2** *The load voltage and the voltage across the load resistance for the half-wave rectifier with an L-R load*

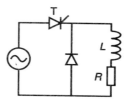

**Figure 9.3** *A half-wave rectifier with a freewheeling diode*

$$Z = \sqrt{R^2 + (\omega L)^2}$$

$$\tan\phi = \frac{\omega L}{R}$$

This expression is only valid for $i$ greater than zero, since the thyristor can only conduct in the forward direction. The current extinction angle, $\beta$, may be found by substituting $\beta$ for $\omega t$ in equation (9.2), setting $i = 0$, and solving for $\beta$. Unfortunately this gives a transcendental equation which can only be solved numerically.

If a freewheeling diode is connected across the load as shown in Figure 9.3 then the load voltage cannot be negative. It will be clamped, and the inductor current transferred or commutated to the diode. With a freewheeling diode, neglecting the diode forward voltage, the current will decay exponentially after voltage zero, hence it will approach zero asymptotically. So, at least in the ideal case, the current in the load will always be continuous. The voltage across the load and the load current (represented by the voltage across R) are illustrated in Figure 9.4. Once again the two shaded areas are equal.

In this case the thyristor conducts for a known phase angle each cycle, from $\alpha$ to $\pi$. This phase angle is referred to as the conduction angle. During the time the thyristor conducts the load voltage equals the supply voltage, for the rest of the cycle it is zero (neglecting the diode forward voltage and the thyristor on-state voltage). Hence the mean load voltage, $V_{av}$, is given by

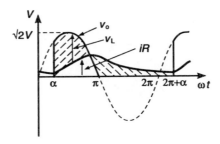

**Figure 9.4** *The load voltage and the voltage across the load resistance for the half-wave rectifier with a freewheeling diode*

$$V_{av} = \frac{\int_{\alpha}^{\pi} \sqrt{2}V\sin\theta \, d\theta}{\int_0^{2\pi} d\theta}$$

(9.3)

$$= \frac{\sqrt{2}V(1 + \cos\alpha)}{2\pi}$$

Since the mean voltage across the inductance is zero, the mean load current is given by

$$I_{av} = V_{av}/R$$

(9.4)

The mean load current can therefore be readily found, provided the mean load voltage is known throughout the cycle.

## 9.3 Single-phase controlled bridge rectifiers

The controlled rectifiers described in Section 9.2 are of little practical value, just as the half-wave diode rectifiers are of little use as power rectifiers. In practice, full-wave rectifiers are required that will draw equal power during the positive and negative half-cycles, and draw no d.c. component from the source. Controlled bridge rectifiers can be classified either as fully controlled rectifiers if all the rectifying devices are controlled switches (thyristors) or as half-controlled if half the elements of the bridge are thyristors and half diodes. The two types of bridge have rather different properties.

### 9.3.1 Single-phase half-controlled rectifiers

There are two possible arrangements of the diodes and thyristors in a single-phase half-controlled converter. These are shown in Figure 9.5 and again an inductive load is assumed. Both circuits will, however, produce the same waveform, which is shown in Figure 9.6. The thyristors are turned on during alternate half-cycles at a

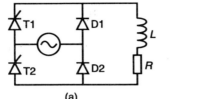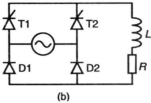

**Figure 9.5** *Two variants on the single-phase half-controlled rectifier*

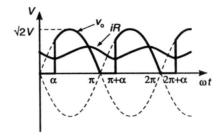

**Figure 9.6** *The load voltage and the voltage across the load resistance for the half-controlled, bridge rectifier with an L-R load*

phase angle $\alpha$ after the voltage zero. One thyristor conducts during the positive half-cycle and one during the negative half-cycle. In both cases the return current is carried by the appropriate diode. For neither circuit can the load voltage become negative, in each case there is a freewheeling path. For the circuit of Figure 9.5(a) the freewheeling path is through the diodes D1 and D2 in series. For the circuit of Figure 9.5(b) the freewheeling path is slightly less obvious. It is through the thyristor that was triggered last and the corresponding diode, T1 and D1 or T2 and D2. In both cases the end result is the same voltage waveform.

Neglecting the forward voltage drop across the thyristors and diodes, the mean load voltage, $V_{av}$, is given by

$$V_{av} = \frac{\sqrt{2}V}{2\pi} \left( \int_{\alpha}^{\pi} \sin \omega t \; d(\omega t) - \int_{\alpha+\pi}^{2\pi} \sin \omega t \; d(\omega t) \right)$$

$$(9.5)$$

$$= \frac{\sqrt{2}V(1 + \cos \alpha)}{\pi}$$

Hence the mean load current may be found, as above, by dividing the mean load voltage by the load resistance. For an inductive load, neglecting the forward voltage of the thyristors and diodes, the load current will always be continuous with a half-controlled rectifier. The mean voltage is controlled from 0 up to $0.9V$ by varying $\alpha$ between 0 and $\pi$.

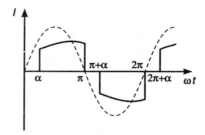

**Figure 9.7** *The source current for the half-controlled converter*

The current flowing on the a.c. side of the rectifier will be as illustrated in Figure 9.7. There are current pulses, of opposite sign, lasting from phase angle $\alpha$ to $\pi$, and from $\alpha + \pi$ to $2\pi$. If the inductance of the load is sufficiently large the load current will be almost constant and free of ripple. In this case the current pulses on the a.c. side of the rectifier will be rectangular.

### 9.3.2 Single-phase fully controlled rectifiers

The fully controlled rectifier differs from the half-controlled rectifier in that all the rectifying devices are thyristors. The circuit of a single-phase fully controlled bridge rectifier is shown in Figure 9.8. There is now no freewheeling path to prevent the load voltage becoming negative. With a sufficiently inductive load the current may be continuous, although as the firing angle $\alpha$ is increased, the load current falls, and for some value of $\alpha$ the current in the load will become discontinuous.

As for the simple half-wave controlled rectifier, the analysis of the circuit can readily be performed to find the load current. Provided the current is discontinuous, the solution for the current is exactly as for the half-wave controlled rectifier with no freewheeling diode, except that there are now two pulses of current per cycle. As before, it is not possible to obtain an analytic solution for the current extinction angle $\beta$. The load voltage and current with discontinuous load current are shown in Figure 9.9.

For larger load current the current through one pair of thyristors will not fall to zero before the second pair are fired and take over the current. The voltage and current waveforms for continuous current are shown in Figure 9.10.

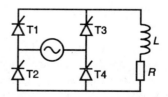

**Figure 9.8** *The circuit of a single-phase fully controlled, bridge rectifier*

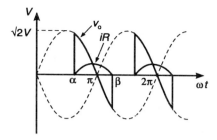

**Figure 9.9** *The load voltage and the voltage across the load resistance of the fully controlled converter with discontinuous current and an L-R load*

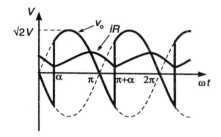

**Figure 9.10** *The load voltage and voltage across the load resistance of the fully controlled converter with continuous current*

When the current is continuous the mean load voltage is given by

$$V_{av} = \frac{\sqrt{2}V}{\pi} \int_{\alpha}^{\pi+\alpha} \sin \omega t \, d(\omega t)$$

$$= \frac{2\sqrt{2}V}{\pi} \cos \alpha$$

(9.6)

The mean load current is again found by dividing the mean load voltage by the load resistance. The mean voltage varies more rapidly with the firing angle, $\alpha$, than is the case for the half-controlled rectifier. This is because as the angle is increased the voltage becomes negative for part of the cycle, while for the half-controlled rectifier it could only fall to zero.

The a.c. current flowing into the rectifier will have a waveform as illustrated in Figure 9.11. If the inductance is sufficiently large this will approximate a square wave, delayed in phase by an angle $\alpha$.

### 9.3.3 Controlled rectifiers with an inductive load and an e.m.f.

Controlled rectifiers are often used to provide a variable d.c. for variable speed d.c. motors. Such a load may be represented by an inductance, a resistance and an e.m.f.

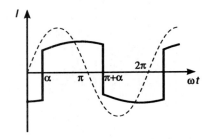

**Figure 9.11** *The source current for the fully controlled converter with continuous load current*

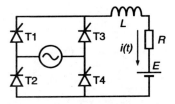

**Figure 9.12** *A single-phase fully controlled bridge rectifier with an L-R-e.m.f. load*

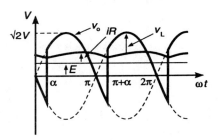

**Figure 9.13** *The load voltage and voltage across the e.m.f. and resistor for the fully controlled rectifier*

in series. Another example of such a load may occur in a battery charger where a controlled rectifier might be used to control the charge rate and an inductance would be needed to smooth the charge current. With such a load the current may be continuous or discontinuous for either a fully controlled or half-controlled rectifier.

The interesting case is that of the fully controlled rectifier with continuous current. The circuit to be considered is shown in Figure 9.12. The load voltage is shown in Figure 9.13, together with how it is distributed between the e.m.f. $E$, the voltage drop across the resistance, and the voltage across the inductance.

The mean load voltage is given by equation (9.6) but now the load current is given by the difference between the load voltage and the e.m.f. divided by the resistance, i.e. the mean load current $I_{av}$ is given by

$$I_{av} = \frac{1}{R} \left( \frac{2\sqrt{2}V}{\pi} \cos \alpha - E \right) \qquad (9.7)$$

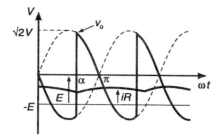

**Figure 9.14** *The load voltage and the voltage across the e.m.f. and the load resistor for the fully controlled converter, when operating as an inverter*

Clearly, $I_{av}$ must be positive, since the thyristors can conduct only in one direction. Normally this will be the result of maintaining the term in cos $\alpha$ greater than the e.m.f., $E$. In this case power is being supplied to the source of e.m.f., and hence drawn from the a.c. supply.

There is, however, another way in which a positive value of $I_{av}$ may be obtained in equation (9.7), that is to make $\alpha$ greater than $\pi/2$, making the first term negative, and to make $E$ negative. In this case the e.m.f. will be supplying power, and power will be returned to the a.c. supply. Thus a circuit which started as a rectifier has become an inverter. The voltage waveforms when the circuit is operating as an inverter are shown in Figure 9.14.

This type of load-commutated converter finds application at very high power levels in high-voltage d.c. transmission systems, such as that which transmits power between France and England. The same power devices and circuits are able to operate as both inverter and rectifier, providing a bi-directional d.c. connection between two a.c. power networks.

### EXAMPLE 9.1

A battery charger uses a single-phase half-controlled rectifier with a source voltage of 110 V at a frequency of 50 Hz. The battery voltage when on charge is 54 V. There is a resistance of 2 $\Omega$ and an inductance of 0.1 H in series with the rectifier and battery. What should be the firing angle, $\alpha$, if the mean charging current is to be 10 A?

### SOLUTION

If the mean charging current is 10 A, the mean load voltage must be 54 + 2×10 = 74 V. The mean load voltage is given by

$$V_{av} = \frac{\sqrt{2} \times 110(1 + \cos \alpha)}{\pi}$$

Thus the firing angle, $\alpha$, is given by

$$\alpha = \cos^{-1}\left(\frac{74\pi}{\sqrt{2} \times 110} - 1\right) = 1.054 \text{ rad}$$

## EXAMPLE 9.2

For the battery charger described in Example 9.1 estimate the ripple current, when the charging current is 10 A.

## SOLUTION

The fundamental ripple frequency will be at $2\omega$, where $\omega$ is the angular frequency of the source ($100\pi$, at a frequency 50 Hz). The impedance of the resistance and the inductance at the fundamental ripple frequency will be given by

$$Z = \sqrt{2^2 + (200\pi \times 0.1)^2} = 62.9\ \Omega$$

The impedance will be greater at higher frequency, so the ripple current will be dominated by the fundamental frequency component. A rough estimate of the ripple voltage across this impedance may be made by observing that the peak-to-peak amplitude of the ripple voltage will be equal to the peak a.c. voltage ($\alpha$ is less than $\pi/2$). Making the rather crude assumption that the amplitude of the fundamental component of the ripple voltage is equal to half the peak a.c. voltage, the amplitude of the ripple current will be given approximately by

$$I_R \approx \frac{110\sqrt{2}}{2 \times 62.9} = 1.24\ \text{A}$$

A better estimate may be found by using Fourier analysis to calculate the amplitude of the ripple voltage at a frequency of $2\omega$. Both the cosine $C_2$ and sine $S_2$ coefficients of the Fourier series will be non-zero, and are given by

$$C_2 = 2\frac{\sqrt{2}V}{\pi} \int_{\alpha}^{\pi} \sin \omega t \cos 2(\omega t)\ \mathrm{d}(\omega t)$$

$$= \frac{2\sqrt{2}V}{3\pi} [3 \cos \theta - 2 \cos^3 \theta]_{\alpha}^{\pi}$$

and

$$S_2 = 2\frac{\sqrt{2}V}{\pi} \int_{\alpha}^{\pi} \sin \omega t \sin 2(\omega t)\ \mathrm{d}(\omega t)$$

$$= \frac{4\sqrt{2}V}{3\pi} [\sin^3 \theta]_{\alpha}^{\pi}$$

where $V$ is the r.m.s. source voltage, $\theta = \omega t$, and use has been made of the fact that the load voltage has a period that is half the period of the source voltage. This means that the integration need only be performed over the interval 0 to $\pi$, and then the result doubled to obtain the integral over the period of the source voltage.

Substituting 110 V for $V$ and 1.054 rad for $\alpha$ gives a value of $-74.0$ for $C_2$ and $-43.3$ for $S_2$. The amplitude of the ripple voltage is given by

$$V_R = \sqrt{C_2^2 + S_2^2} = 85.8 \text{ V}$$

Hence the amplitude of the ripple current is given by 85.8/62.9, or 1.36 A. This is in reasonable agreement with the value obtained by the previous simple estimate.

## 9.4 The power factor of single-phase controlled rectifiers

The power factor of simple diode bridge rectifiers was considered in Chapter 4, Section 4.7 for both a simple capacitance filter and an inductive filter. The same concept will now be applied to the controlled rectifier.

For a single-phase controlled rectifier with a sufficiently large load inductance the load current can be considered to be constant. This condition will be met if the load time constant is very much greater than the period of the a.c. source, say, ten or twenty times the period. The current on the a.c. side of the rectifier will also be constant while the thyristors are conducting, but the polarity will change depending on which half of the bridge is in conduction. The source currents for the half- and fully controlled converters will be as shown in Figures 9.7 and 9.11, but the positive and negative current pulses will be rectangular with amplitude $I_0$, where $I_0$ is the (constant) load current.

For the half-controlled rectifier the thyristors will conduct for a phase angle between $\alpha$ and $\pi$, during each half-cycle. The r.m.s. source current will therefore be given by

$$I_{\text{rms}} = \sqrt{\frac{\int_\alpha^\pi I_0^2 d\theta}{\pi}} = I_0 \sqrt{\frac{\pi - \alpha}{\pi}} \tag{9.8}$$

The power factor is the ratio of the true power to the product of the r.m.s. current and the r.m.s. voltage. If the power loss in the thyristors and diodes is ignored, then the power drawn from the source is equal to the power dissipated in the load, i.e. the load voltage multiplied by $I_0$. Hence the power factor, PF, for a single-phase half-controlled rectifier is given by

$$\text{PF} = \frac{\sqrt{2}V(1 + \cos\alpha)I_0}{\pi} \cdot \frac{1}{VI_0}\sqrt{\frac{\pi}{(\pi - \alpha)}}$$

$$= \frac{\sqrt{2}(1 + \cos\alpha)}{\sqrt{\pi(\pi - \alpha)}}$$

$$\tag{9.9}$$

For the fully controlled rectifier the source current is a square wave of amplitude $I_0$, hence the r.m.s. of the source current is $I_0$. Dividing the load power by the VA product of the a.c., the power factor for a single-phase fully controlled rectifier is given by

$$\text{PF} = \frac{2\sqrt{2}VI_0 \cos \alpha}{\pi VI_0} = \frac{2\sqrt{2} \cos \alpha}{\pi} \qquad (9.10)$$

The power factor is less than unity for both types of rectifier for two reasons. The source current is not sinusoidal, and while the harmonic current components contribute to the r.m.s. current, they do not contribute to the power transfer. The second effect is that there is a phase shift between the fundamental current and the source voltage. For the fully controlled rectifier the phase of the current lags the voltage by an angle $\alpha$, while for the half-controlled converter the phase lag is $\frac{1}{2}\alpha$. This can be seen by inspecting the current waveforms in Figures 9.7 and 9.11, remembering that the load current is constant and hence the waveforms will be rectangular in shape.

In order to characterize the relative importance of the phase shift and the harmonic distortion it is convenient to define two other figures of merit, the *harmonic factor* (HF) and the *displacement power factor* (DPF). The harmonic factor is the ratio of the r.m.s. value of the current at the line frequency $I_{1(\text{rms})}$ to the r.m.s. current $I_{(\text{rms})}$. While the displacement power factor is $\cos \phi_1$, where $\phi_1$ is the phase angle between the line voltage and the fundamental component of the line current. Using equation (4.27), the power factor is related to the harmonic factor and displacement power factor by

$$\text{PF} = \frac{I_{1(\text{rms})}\cos \phi_1}{I_{(\text{rms})}} = \text{HF.DPF} \qquad (9.11)$$

It has been assumed that the supply voltage is sinusoidal.

For a constant load current, the harmonic content of the source current does not change with firing angle for the fully controlled rectifier. Since the shape of the waveform remains constant, only the phase lag and amplitude change. Since the phase shift of the input current is $\alpha$, from equation (9.10) it can be seen that the harmonic factor is $2\sqrt{2}/\pi$, while the displacement power factor is $\cos \alpha$.

However, for the half-controlled rectifier the shape of the waveform, and its harmonic content, also change. The displacement power factor is given by $\cos \frac{1}{2}\alpha$, and equation (9.9) can be rewritten in the form

$$\text{PF} = \frac{2\sqrt{2} \cos^2 \frac{1}{2}\alpha}{\sqrt{\pi(\pi - \alpha)}}$$

from which it may be seen that the harmonic factor is given by

$$\text{HF} = \frac{2\sqrt{2} \cos \frac{1}{2}\alpha}{\sqrt{\pi(\pi - \alpha)}} \qquad (9.12)$$

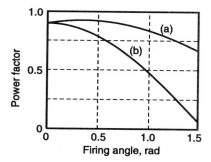

**Figure 9.15** *The power factor for (a) a half-controlled rectifier and (b) a fully controlled rectifier, with a constant load current*

For small values of $\alpha$ the total harmonic content decreases with increasing $\alpha$, and the power factor improves. This is illustrated in Figure 9.15, where the power factors for both the half- and fully controlled rectifiers are shown. Note, however, this assumes that the load current is continuous, and that the ripple is negligible. These assumptions may break down for large values of $\alpha$.

## 9.5 Commutation in controlled rectifiers

With an uncontrolled single-phase bridge rectifier, operating with continuous current in an inductive load, the commutation of current between the diodes takes place at the voltage zero of the source waveform. For a controlled rectifier, commutation takes place when the source voltage is non-zero. However, in both cases, during commutation all the elements of the bridge will be in a conducting state, and the source is essentially shorted, until the current in one pair of devices falls to zero.

The source will, in practice, always have a finite inductance, from the leakage inductance of the transformer or just the inductance of the wires. Sometimes *line inductors* are inserted in series with the power source to limit the rate of change of current and protect the thyristors. In order to understand what happens during commutation, consider the fully controlled bridge rectifier with a source impedance, $L_S$, as shown in Figure 9.16.

Assume that the load current $I_0$ is constant and that the source voltage, $v_S$, is positive. The thyristors T2 and T3 have been conducting and the current is now being commutated to T1 and T4. All devices in the bridge are conducting, so the bridge voltage will be zero. The current in the inductor must change from $-I_0$ to $+I_0$. Since the bridge voltage is zero, the source current during commutation will obey the differential equation

$$\sqrt{2}V \sin \omega t = L_S \frac{di_S(t)}{dt} \tag{9.13}$$

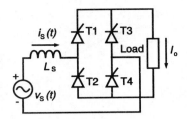

**Figure 9.16** *A fully controlled rectifier with line inductance*

Integrating the equation with respect to $t$, from the start of the commutation at phase angle $\alpha$ to its completion at phase angle $\alpha + \gamma$, gives

$$\sqrt{2}V \int_{\alpha}^{\alpha+\gamma} \sin \omega t\, \mathrm{d}(\omega t) = \omega L_S \int_{-I_0}^{I_0} \mathrm{d}(i_S)$$

hence,

$$\cos(\alpha + \gamma) = \cos \alpha - \frac{\sqrt{2}\omega L_S I_0}{V} \tag{9.14}$$

Provided that $\gamma$ is small and $\sin \alpha$ is not too small, equation (9.14) may be approximated to give

$$\gamma \approx \frac{\sqrt{2}\omega L_S I_0}{V \sin \alpha} \tag{9.15}$$

The effect of commutation on the load voltage waveform is shown in Figure 9.17. The duration of the commutation *notch* during which the load voltage is zero depends on the source inductance and on the firing angle, $\alpha$. The angle $\gamma$ is referred to as the overlap angle.

## 9.6 Simulation of controlled rectifiers

One of the difficulties that arises when attempting to simulate controlled rectifiers

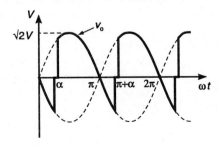

**Figure 9.17** *The load voltage of a fully controlled rectifier with line inductance, showing the commutation notch*

is that SPICE has no model for a thyristor. It is necessary to devise a *macro-model* or sub-circuit which models the behaviour of the thyristor devices. Rashid (1993) and Mohan, Undeland and Robbins (1995) give models for thyristors, based on the use of voltage-controlled switches. The evaluation version of PSPICE has a comprehensive, but complex thyristor model in the evaluation library, based on switches and the analog modelling facilities of PSPICE. The basis of these models is that the switch is controlled by a non-linear function of the gate voltage and the current through the device. The difficulty that arises is that the large non-linearity, and the positive feedback, may cause instability in the integration routine of the simulator, which will then fail because the step size becomes smaller than the minimum allowed value.

Another, simpler, approach which may be adequate is to represent the thyristor by a diode and a switch in series. The switch is controlled by a voltage timed to turn it on at the correct time, and to keep it on until after the current in the diode has fallen to zero. This is obviously possible only if the time is known in advance, at which the switch may be reopened , without breaking the current, but before the diode is again forward biased. Provided the operation of the converter is understood, it should usually be possible for this to be arranged. This simple approach has been adopted to model the converter described in Example 9.1.

### 9.6.1 Simulation of a half-controlled rectifier

The SPICE listing for the rectifier is given in Appendix 1, Section A1.6. The simple diode and switch model for the thyristor has been included as the subcircuit SCR. Note that the syntax for the switch device is that appropriate to PSPICE. Where other versions of SPICE are used the syntax for a switch (if available) may be different. An RC snubber is connected across the switch to avoid spikes as the thyristors are switched. The parameter *delay* controls the firing angle, while the parameter *switch-time* controls the time each switch is held on. The time chosen is such as to ensure that the switch is on until after the diode in the model ceases conduction. The rectifier freewheels through the diodes D1 and D2. The initial condition for the current in inductor LL has been chosen, by trial and error, to give the steady-state load condition.

The firing angle was chosen as 1.054 rad, the value for which the simple analysis gave a mean load current of 10 A. The simulated waveform of the voltage across the load is shown in Figure 9.18 and the load current in Figure 9.19. The mean load current is obtained from Fourier analysis of the load current (the .four statement). This gave a mean current of 8.84 A. This is less than the 10 A predicted by the simple analysis, which ignored the voltage drop across the diodes and thyristors. The fundamental component of the ripple current was 1.34 A, in good agreement with the value predicted from the simple analysis (1.36 A).

The simulated current from the a.c. source is shown in Figure 9.20. The waveform is not flat topped, since the load current has a significant ripple component. The r.m.s. source current deduced from the Fourier analysis is about

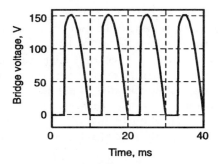

**Figure 9.18** *The load voltage for the simulated half-controlled rectifier with an L-R-e.m.f. load*

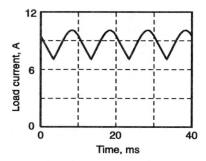

**Figure 9.19** *The load current for the simulated half-controlled rectifier with an L-R-e.m.f. load*

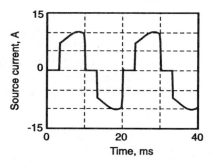

**Figure 9.20** *The line current for the simulated half-controlled rectifier with an L-R-e.m.f. load*

7.54 A, while the fundamental component has an r.m.s. magnitude of about 7.35 A, hence the harmonic factor is about 0.975. That the value is so high is largely due to the suppression of the third harmonic, a point which will be considered in more detail in Chapter 10. The phase delay of the fundamental is 0.597 rad, rather greater than would be predicted assuming a rectangular waveform. This gives a displacement power factor of 0.827. The overall power factor is therefore 0.806.

## Summary

This chapter has provided an introduction to controlled rectification. For simplicity, only single-phase circuits have been considered, although in practice controlled rectification is used most often at power levels of tens of kilowatts or more, where three-phase circuits would most probably be used. The way in which a controlled rectifier controls the load voltage has some similarity with a buck regulator in that the load voltage is a series of pulses, although the pulses are not of a simple rectangular shape. The steady-state transfer characteristics of controlled rectifiers with continuous current in the load have been derived. When the current becomes discontinuous analysis becomes more complex and exact analytic solutions are not possible. The power factor presented by a controlled rectifier has been considered. As with all rectifiers, this may be poor, and, especially at high power levels, careful filtering is needed to remove high-frequency components from the supply line.

## Self-assessment questions

9.1 A single-phase full-wave half-controlled bridge rectifier with an L-R load has a source voltage of 250 V r.m.s. The required load mean load voltage is 150 V d.c. What will be the firing angle $\alpha$ for this load voltage?

9.2 A single-phase fully controlled bridge rectifier has a source voltage of 200 V r.m.s. at a frequency of 50 Hz. The load has an inductance of 1 H and a resistance of 10 $\Omega$. If the firing angle, $\alpha$, is 60°, what is the mean load current, assuming that the load current is continuous? Is this assumption reasonable?

9.3 A single-phase full-wave half-controlled rectifier has a source voltage of 115 V at a frequency of 60 Hz. The firing angle, $\alpha$, is 0.5 rad, and the load has a resistance 20 $\Omega$ and an inductance of 100 mH. Estimate the amplitude of the ripple current.

9.4 A single-phase half-controlled rectifier operates with a constant load current. If $\alpha$ is 1 rad, find the value of the power factor and the harmonic factor.

9.5 A single-phase fully controlled rectifier operates with a firing angle 0.5 rad. The source has an r.m.s. voltage of 240 V at 50 Hz and a source inductance of 1 mH. The load current is 10 A, and approximately constant. What is the duration of the commutation notch, expressed (a) as a phase angle and (b) in seconds?

## Tutorial questions

**9.1** Sketch the voltage waveform for a fully controlled single-phase bridge rectifier with continuous current flowing in the load.

**9.2** For the rectifier of Question 9.1, sketch the current flowing in the a.c. line, assuming that the load current is constant.

**9.3** Sketch the circuits for the two forms of the half-controlled rectifier, and describe the freewheeling path in each case.

**9.4** For a half-controlled rectifier with a constant load current sketch the load voltage and source current waveforms.

**9.5** Discuss how a fully controlled bridge rectifier may also operate as an inverter.

**9.6** Explain the concept of power factor, displacement power factor, and harmonic factor.

**9.7** Explain the origin of the commutation notch.

## References

Davis, R. M., *Power Diode and Thyristor Circuits*, Cambridge University Press, Cambridge, 1971.

Lander, C. W., *Power Electronics*, 3rd edition, McGraw-Hill, Maidenhead, 1993.

Mohan, N., Undeland T. M. and Robbins, W. P., *Power Electronics: Converters, Applications and Design*, John Wiley, New York, 1995.

Rashid, M. H., *Power Electronics: Circuits, Devices, and Applications*, 2nd edition, Prentice Hall, Hemel Hempstead, 1993.

CHAPTER 10

# *Power inversion*

## 10.1 Introduction

Power inversion is the process of converting d.c. electrical power to a.c. There is a wide range of applications for this type of power conversion, ranging in power levels from a few tens of watts for providing portable sources of a.c. power for domestic applications to many megawatts in industrial and power system applications. The active power devices and techniques used obviously vary with application and power level, but the principles remain the same. In this chapter only single-phase inverters will be considered, although the same techniques can readily be applied to three-phase inverters.

An inverter may be required either to supply power to an independent load, which is supplied only from the inverter, or power to a network which is also supplied by other sources. In the first case the frequency and voltage of the output are determined by the inverter. An example of such an application would be the generation of a.c. from a storage battery to provide power when connection to the public electricity supply is not possible, or during failure of the main supply. In the second case the load frequency and voltage may, to a large extent, be fixed by the other sources connected to the network, and to connect the inverter safely, it must be synchronized to the network. An example of this type of application would be the supply of power from a photovoltaic solar power generator to the local power distribution network. Obviously the requirements are rather different for these two scenarios. In the first case an inverter that behaves as a voltage source with low output impedance is probably required, while in the second case it might be easier to control the inverter if it behaved like a current source.

## 10.2 Voltage-source inverters

The simplest form of single-phase inverter uses a half- or full-bridge arrangement of switches to switch the polarity of the voltage across the load. The basic circuits are shown in Figure 10.1. For a voltage-source inverter the d.c. source is assumed to be an ideal voltage source with a low output impedance. For the half-bridge inverter a centre-tapped voltage supply is needed (Figure 10.1(a)). At low power it may be possible to use a simple supply with two well-matched capacitors in series

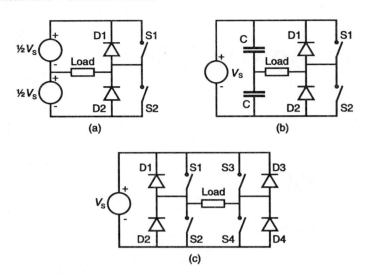

**Figure 10.1** *Single-phase voltage-source inverters. (a) Half-bridge; (b) half-bridge with capacitive centre tap; (c) full bridge*

to provide the centre tap (Figure 10.1(b)). However, the capacitors will need to have a large value to provide a low impedance at the switching frequency. The bridge inverter (Figure 10.1(c)) avoids the need for a tapped supply, and gives twice the peak output voltage for the same total supply voltage.

In the simplest mode of operation the switches are operated in pairs to produce a square wave load voltage. For the half-bridge converter S1 is turned on and S2 off to give a positive load voltage of $\frac{1}{2}V_S$, and S2 is turned on and S1 turned off to give a voltage of $-\frac{1}{2}V_S$. For the full-bridge converter, S1 and S4 are on and S2 and S3 are off to give a load voltage of $V_S$ and S2 and S3 are on and S1 and S2 off to give a load voltage of $-V_S$.    S4

If the load is resistive the current flowing in the load reverses as soon as the switches reverse the polarity of the load voltage. However, if the load is inductive then the current cannot immediately reverse its direction. Consider the case of a half-bridge inverter with inductive load (Figure 10.2). S2 has been conducting and has started to open and S1 has started to close. As soon as S2 starts to open and its current starts to fall the potential across S2 will rise rapidly. If S1 has not yet turned on, or is unable to carry the load current in the reverse direction, then the voltage across S2 will rise to above the source voltage and D1 will conduct. The current in the load will fall until D1 turns off and the current is commutated to S1. When S1 turns off a similar process occurs with the current commutated first to D2 then to S2. The voltage and current in the load are shown in Figure 10.3. This figure also indicates which of the devices is carrying the current during each part of the cycle.

The supply current is positive while the switches are carrying the load current, but while the freewheeling diodes are conducting the current flows in the reverse direction, returning energy to the source. The symmetry of the load waveform

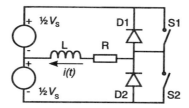

**Figure 10.2**  *A half-bridge inverter with an L-R load*

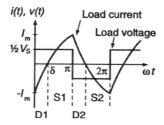

**Figure 10.3**  *The load voltage and current waveform of a half-bridge inverter with an L-R load*

requires that, with an inductive load, the peak reverse current is equal to the peak forward current. For the waveform in Figure 10.3 the current flowing in the source is as shown in Figure 10.4. When the load is highly inductive the amplitude of the alternating component of the source current may well exceed the d.c. component of the current. With a load that is a pure inductance there is no d.c. component, since no power is delivered to the load. Clearly, when the inverter supplies power to a reactive load the voltage source must be able to source and sink current. If it cannot do so then it must be shunted by a reservoir capacitor with a value sufficiently large that it will bypass the circulating a.c. component, without significant voltage excursions.

In the above description it has been assumed that the main switches are unidirectional devices. This may be true if the devices are thyristors with forced commutation, but many power switches will conduct in the reverse direction. For example, power MOSFETs have a built-in reverse diode which would carry the reverse current. However, it is generally better to use an external diode to carry the reverse curent. If it is necessary to prevent any reverse current flowing in the switching device, then a series diode may be included to force the current through the parallel diode (Figure 10.5).

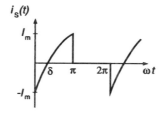

**Figure 10.4**  *The current in the positive supply for the half-bridge converter with an L-R load*

**Figure 10.5** *Preventing current in the reverse diode of a power MOSFET*

For the full-bridge inverter the diodes operate in the same way as for the half-bridge inverter, carrying the current flowing back into the source. The diodes also provide a path for the current to flow if the switch turning on is delayed slightly after the switch turning off. This slight delay is obviously desirable to avoid the possibility of both switches being on simultaneously and shorting the power supply.

### EXAMPLE 10.1
A full-bridge inverter produces a square wave output voltage with an amplitude of 100 V at a frequency of 100 Hz. The load connected to the inverter is equivalent to a resistance of 10 Ω in series with an inductance of 30 mH. Find the peak value of the load current. For what fraction of each half-cycle is the load current carried by the freewheeling diode?

### SOLUTION
While the load voltage is positive the equation for the current is

$$0.03 \frac{di}{dt} + 10i = 100$$

The general solution is of the form

$$i = 10 + Ae^{-t/\tau}$$

where $\tau = 0.003$ and $A$ is a constant. The initial condition at the start of the half-cycle is $i = -I_m$, hence

$$A = -I_m - 10$$

and

$$i = 10 - (10 + I_m)e^{-t/\tau}$$

At the end of the half-cycle the current must equal $I_m$, since the waveform is symmetric, hence

$$I_m = 10 - (10 + I_m)e^{-\frac{T}{2\tau}}$$

or

$$I_m = \frac{10(1 - e^{-\frac{T}{2\tau}})}{1 + e^{-\frac{T}{2\tau}}}$$

where $T$ is the period (10 ms). Substituting for 0.003 for $\tau$, and 0.01 for $T$ gives $I_m = 6.82$. Hence the peak load current is 6.82 A.

Setting $i$ equal to zero gives the time at which the current is commutated from the freewheeling diode to the switch:

$$0 = 10 - 16.82e^{-t/\tau}$$

or

$$t = -\tau \ln(10/16.82)$$

Substituting for $\tau$ gives $t = 1.56 \times 10^{-3}$. The duration of the half period is 5 ms, hence the freewheeling diode conducts for a fraction 1.56/5, or 0.312 of the half-cycle.

## 10.3 Current-source inverters

The voltage-source inverters described in Section 10.2 cannot operate with a capacitive load. When the polarity changes the current in the switches is not limited and large current spikes may occur, limited only by the source impedance and the rate of rise of current in the switches as they turn on. In a similar way, if a voltage-source inverter is used to supply power to a busbar energized from another source, with low impedance, then again there is no means of limiting or controlling the current.

Where current limitation and control is required, the voltage source on the d.c. side of the inverter may be replaced with a current source. In practice this is achieved by the use of a large inductance to hold the current constant on the timescale of the switching, with some means of controlling the voltage feeding the inductance (Figure 10.6). The variable voltage source may be a switching regulator for efficiency.

The current-source inverter shown in Figure 10.6 may be connected to a resistive or a capacitive load, or to a load which is voltage stiff, i.e. the load voltage

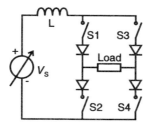

**Figure 10.6** *A current-source inverter*

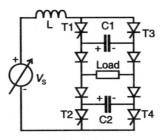

**Figure 10.7** *A current-source inverter using thyristor switches, with forced commutation*

is determined by a second power source which controls the voltage waveform. If the load is inductive large voltage spikes will occur as the direction of the load current is forced to reverse. The magnitude of such spikes depends on the rate at which the current is commutated between the switches and hence reversed. The diodes in series with the switches are to protect the switches from reverse voltages.

One feature of this circuit is that it is readily adapted to the use of thyristor switches, and is therefore suitable for use at high power. The transistor switches in Figure 10.6 are replaced by thyristors and capacitors added to force commutation (Figure 10.7). To see how the current is commutated, consider the case when T1 and T4 are in conduction. Capacitors C1 and C2 are both charged with a positive voltage as shown. T3 and T2 are both forward biased, and hence can be triggered. When T3 and T2 are fired the charged capacitors will apply a reverse potential across T1 and T4, forcing them to turn off. During the negative half-cycle the capacitors will recharge with the opposite polarity, ready to force commutation when T1 and T4 are fired.

Another current-source inverter which is particularly well suited for use in very high power d.c. interconnections in power systems is the load-commutated fully controlled thyristor bridge (Figure 10.8). This is based on the fully controlled bridge rectifier and was discussed briefly in Section 9.3.3. The commutation of the thyristors in this circuit depends on the load-side voltage. It can therefore normally only be used to supply extra power to an existing a.c. supply (or possibly drive a parallel resonant LC load). For high-power applications it would normally be used in a three-phase version. For simplicity only the single-phase inverter is considered here.

The circuit of the load commutated converter (Figure 10.8) has been drawn to emphasize its similarity to the basic current-source inverter (Figure 10.6) rather

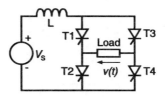

**Figure 10.8** *A load-commutated current-source thyristor inverter*

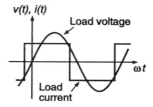

**Figure 10.9** *The load voltage and current for a load-commutated thyristor inverter*

than the way it was drawn in Chapter 9, where it was primarily considered as a rectifier. The load voltage and load current waveforms are shown in Figure 10.9, assuming that the source current is constant. The load current must lead the load voltage, otherwise commutation is not possible. The lead angle should generally be as small as possible, while ensuring reliable commutation so as to minimize the reactive power that is supplied by the load.

## 10.4 Control of inverters

The inverters considered above produce a square voltage (or current) waveform with an amplitude determined by the source voltage (or current). Obviously one way of controlling the load voltage (or current) is to control the source voltage (or current). This could be done using a d.c.–d.c. converter with pulse-width or pulse-frequency modulation. However, the two functions can be combined by applying the pulse-width modulation directly to the inverter.

For a voltage-source full-bridge inverter the load voltage may be positive, negative or zero, depending on which pair of the four switches are on. If the upper switch on one side of the load and the lower switch on the other side are on, the output voltage will be either positive or negative. If two switches in either the upper or the lower halves of the bridge are on, i.e. S1 and S3 or S2 and S4, then the load voltage will be zero. When the load voltage is zero, any reactive current will flow in one of the switches, and in the reverse diode across the other switch, depending on the direction of the load current. Where the voltage (current) is switched between a positive value and zero, or a negative value and zero, the modulation is referred to as unipolar PWM.

For the half-bridge inverter the load voltage cannot be zero. Pulse-width modulation still can be used, but the load voltage can be switched only between positive and negative, and this type of modulation is referred to as bipolar PWM. A full-bridge inverter can, of course, also produce bipolar PWM.

For current-source inverters pulse-width modulation can be used by diverting the source current from the load. With reference to Figure 10.6, this involves closing S1 and S2 or S3 and S4. This will give unipolar modulation of the load current. Alternatively, the load current may be pulse-width modulated by reversing the direction of the current flow (bipolar PWM). For a load-commutated converter pulse-width modulation of the inverter is clearly not possible.

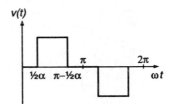

**Figure 10.10**  *Single-pulse modulation of a voltage-source inverter*

In the sections which follow voltage control of single-phase voltage-source inverters will be considered. It will also be assumed that the inverter is of the full-bridge type, permitting unipolar or bipolar modulation. The same techniques can be used to control the load current of a current-source inverter.

### 10.4.1 Single-pulse modulation

The simplest method of controlling the load voltage of a single-phase inverter is to modify the square wave output by introducing a period of zero voltage with a phase angle $\alpha$ between the positive and negative parts of the cycle. The waveform is as shown in Figure 10.10.

The value of the r.m.s. voltage of the waveform shown in Figure 10.10 is given by

$$
V_{\text{rms}} = \sqrt{\frac{\int_{\alpha/2}^{\pi-\alpha/2} V_S^2 \, \mathrm{d}(\omega t)}{\pi}}
$$

$$
= V_S \sqrt{\frac{\pi - \alpha}{\pi}}
$$

(10.1)

Clearly as $\alpha$ increases, the r.m.s. of the load voltage falls. The harmonic content of the load voltage may be found by Fourier analysis. As drawn in Figure 10.10, the waveform is an odd function so all the cosine terms in the Fourier series are zero, and the amplitude of the sine component at a frequency $n\omega$ is given by

$$
S_n = \frac{V_S}{\pi} \int_{\alpha/2}^{\pi-\alpha/2} \sin n\omega t \, \mathrm{d}(\omega t)
$$

$$
+ \frac{V_S}{\pi} \int_{\pi+\alpha/2}^{2\pi-\alpha/2} \sin n\omega t \, \mathrm{d}(\omega t)
$$

(10.2)

$$
= \frac{4V_S}{n\pi} \cos n\alpha/2 \quad \text{if } n \text{ is odd}
$$

$$
\text{or} = 0 \quad \text{if } n \text{ is even}
$$

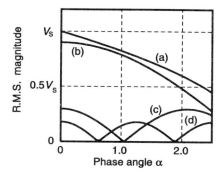

**Figure 10.11** (a) The r.m.s. load voltage and the r.m.s. magnitudes of (b) the fundamental, (c) the third harmonic and (d) the fifth harmonic of the load voltage for single-pulse modulation

and the Fourier series giving the load voltage may be written

$$v(t) = \Sigma_{n=1,3,5,..} \frac{4V_S}{\pi} \cos\left(n\alpha/2\right)\sin\left(n\omega t\right) \tag{10.3}$$

In Figure 10.11 the r.m.s. load voltage, the r.m.s. amplitudes of the fundamental and some of the harmonics are shown . It can be seen that for each harmonic there are values of $\alpha$ for which the amplitude becomes zero. For the $n$th harmonic the condition that the amplitude is zero is that

$$\cos\frac{n\alpha}{2} = 0$$

This property can be used to reduce the total harmonic distortion of the load waveform by eliminating the third harmonic component, although, by doing so control of the load amplitude is lost.

### 10.4.2 Multiple-pulse modulation

In order to avoid large total harmonic distortion at low load voltage the two half-cycles may be modulated at a high frequency with rectangular pulses as shown in Figure 10.12. The load voltage may be controlled by varying the duty ratio of the high-frequency pulses. The frequency spectrum of the resulting waveform may be divided into two parts, the low-frequency spectrum with its fundamental at an angular frequency $\omega$, together with its harmonics, and a high-frequency spectrum related to the switching frequency. The high-frequency components are filtered out relatively easily, leaving only the low-frequency ones, which will follow the spectrum of a square wave.

The pulse-width modulation used in Figure 10.12 is unipolar, with a polarity reversal as the output polarity changes phase. The alternative approach is to use the switches to generate a bipolar pulse train as shown in Figure 10.13. In this case a duty ratio of 0.5 gives a zero output voltage. Higher or lower ratios give positive or

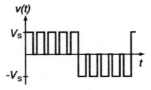

**Figure 10.12**  *Unipolar multiple-pulse modulation*

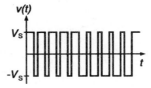

**Figure 10.13**  *Bipolar multiple-pulse modulation*

negative voltages. This approach has the disadvantage that the magnitude of the switching voltage is greater, but a half-bridge inverter can be used.

In a pulse-width modulated inverter it is desirable that the switching frequency for the pulse-width modulation should be an exact multiple of the fundamental a.c. frequency and phase locked to it, otherwise unwanted subharmonics of the fundamental frequency may be generated. For a single-phase bipolar PWM inverter the switching frequency should be an odd multiple of the fundamental frequency, to avoid the generation of even harmonics (of the fundamental frequency). For a unipolar PWM inverter the switching frequency should be an even multiple of the fundamental frequency, with a phase reversal as the load voltage changes sign. This symmetry ensures that all the even harmonics will cancel, including the component at the switching frequency.

### EXAMPLE 10.2

A voltage-source single-phase half-bridge inverter is multiple-pulse modulated with a duty ratio of 0.75 while the fundamental component of the load voltage is positive and 0.25 while it is negative. The supplies to the inverter are +100 V and −100 V. What is (a) the r.m.s. load voltage, (b) the mean load voltage (averaged over one period of the pulse modulation) while the output is positive and while it is negative, and (c) the r.m.s. magnitude of the fundamental component of the load voltage? Assume that the switching frequency is very much greater than the fundamental frequency.

### SOLUTION

(a) The load voltage is always +100 V or −100 V. In finding the r.m.s. the sign is lost when the voltage is squared. Hence when the squared voltage is averaged the value will be $100^2$, and the r.m.s. value will be 100 V.

(b) The mean voltage during the positive half-cycle will be given by 0.75 × 100 − 0.25 × 100 = 50 V and during the negative half-cycle it will be 0.25 × 100 − 0.75 × 100 = −50 V.

(c) If the pulse modulation frequency is very much higher than the fundamental frequency, then the fundamental Fourier component of the output voltage may be deduced by first averaging the waveform over the pulse period to obtain a square waveform of amplitude 50 V and then finding the amplitude of the fundamental component $S_1$. Taking the amplitude of the waveform as +50 for the first half-cycle and −50 for the second half-cycle the amplitude of the fundamental component is given by

$$S_1 = \frac{50}{\pi} \int_0^{\pi} \sin \omega t \, d(\omega t) - \frac{50}{\pi} \int_{\pi}^{2\pi} \sin \omega t \, d(\omega t)$$
$$= \frac{50 \times 4}{\pi} = 63.7$$

Hence the r.m.s. magnitude of the fundamental component of the load voltage is $63.7/\sqrt{2} = 45.0$ V.

From this example it can be seen that for a bipolar PWM the r.m.s. of the load voltage is not reduced by the modulation but the fundamental voltage and the lower harmonics are reduced. Provided the modulation frequency is high, the high-frequency harmonics, which contribute significantly to the r.m.s. value of the load voltage, may be filtered out, leaving only the fundamental and the lower harmonics. For a unipolar PWM the r.m.s. value will also be reduced, since for part of the cycle the load voltage is zero.

### 10.4.3  Sinusoidal pulse-width modulation

Multiple-pulse modulation with a constant duty ratio for the pulses during the positive and negative half-cycles of the load enables the load voltage to be controlled but the low-frequency harmonic content is the same as for a square wave. The harmonic content may be reduced by programming the duty ratio so that the load voltage, averaged across the switching period, follows a sinusoid.

The pulse train may be generated using a pulse-width modulator similar to that described in Chapter 7. For bipolar switching of the inverter a sinusoidal wave is compared with a triangular waveform using a comparator. The output of the comparator controls the switches. If the sinusoid exceeds the amplitude of the triangular wave the output of the inverter is set positive, if not, it is set negative, as shown in Figure 10.14. A triangular carrier waveform is preferred to the sawtooth waveform used in switched mode power supply controllers, since it gives a constant

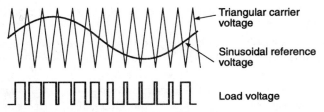

**Figure 10.14** *Generating bipolar sinusoidal pulse-width modulation*

time between successive samples of the reference voltage, even when the reference voltage is changing rapidly. This reduces the distortion of the pulse-width modulated output signal.

For unipolar switching of the inverter a rather more complex sequence of switching is required. One method of achieving the required switching sequence is to use two sine wave reference signals in anti-phase and two comparators. The switching of the inverter bridge is controlled by switching one 'leg' of the bridge (i.e. S1 and S2) from the first comparator, which compares reference P with the triangular carrier and the other leg of the bridge (S3 and S4) from the second comparator, which compares reference Q with the carrier. If reference P, for the first comparator, is less than the triangular carrier, switch S1 is closed and S2 open. If reference P exceeds the carrier, then S2 is closed and S1 open. For the other leg of the bridge switches S3 and S4 are controlled in the same way using reference Q and the second comparator. The waveforms are illustrated in Figure 10.15. Note that the modulation of the output voltage is at twice the frequency of the carrier. The output of the bridge is not left floating, except briefly during the switching transitions, when it is important to avoid short circuiting the source.

The use of analog methods of generating the pulse-width modulation is giving way to methods based on the use of high-speed microcontrollers or micro-

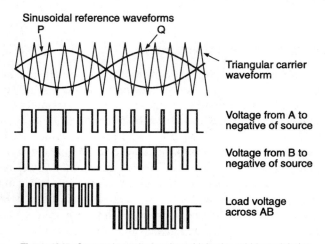

**Figure 10.15** *Generating unipolar sinusoidal pulse-width modulation*

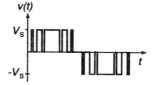

**Figure 10.16** *Modified unipolar sinusoidal pulse-width modulation*

processors. The advantages to be gained are flexibility and a more stable source of modulation, free from problems of drift. However, a high-performance micro-processor with a fast clock speed is required, and probably hardware dedicated to generating the pulses.

### 10.4.4 Modified sinusoidal pulse-width modulation

Sinusoidal PWM enables an inverter to give a low distortion sinusoidal output and a good method of controlling the load voltage. The main disadvantages are that the high switching frequency may result in large switching losses and low efficiency, and that the r.m.s. amplitude of the fundamental component of the output is substantially less than the source voltage. One way of reducing the low-frequency harmonic content while avoiding a very large number of switching cycles per period of the fundamental frequency is to stop the switching during the central part of the positive and negative half-cycles where the voltage of a sine wave is changing only slowly. The resultant waveform is as illustrated in Figure 10.16. This procedure is referred to as *modified sinusoidal pulse-width modulation*. It reduces the switching losses and also increases the amplitude of the fundamental component of the output waveform.

## 10.5 Simulation of a sinusoidal pulse-width modulated inverter

The inverter chosen for simulation is a current-source inverter supplying a load which is 'voltage stiff', i.e. a voltage source. The circuit is shown in Figure 10.17. The switches are simple and voltage-controlled with an on-resistance of 0.1 $\Omega$ and an off resistance of 1 M$\Omega$. The load voltage VL is also used as the sinusoidal reference voltage for the PWM, together with the voltage sources VR1 and VR2, which generate the triangular carrier at the switching frequency. Switches S3 and S2 are on during the positive half-cycle and S1 and S4 during the negative half-cycle. During the positive half-cycle S1 is turned on to divert the current from the load and achieve the pulse modulation. The diodes prevent current flowing through the load. During the negative half-cycle S3 is used to divert the current. The switch control is achieved simply by using the voltage-controlled switches as comparators. A netlist is included in Section A1.7 of Appendix 1.

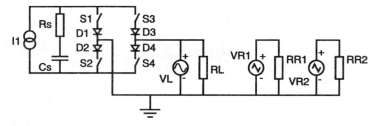

**Figure 10.17** *The circuit for the simulation of a current-source inverter with sinusoidal pulse-width modulation*

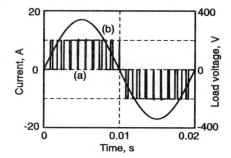

**Figure 10.18** *The load current (a) and the load voltage (b) for the simulated inverter*

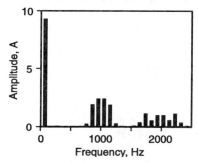

**Figure 10.19** *The frequency spectrum of the load current for the sinusoidal pulse-width modulated inverter*

The load current and load voltage waveforms are given in Figures 10.18 (a) and (b) respectively. The frequency spectrum of the load current is shown in Figure 10.19. The switching frequency for the PWM is 1 kHz, but this component is suppressed, along with all the even harmonic components. This is because the waveform is symmetrical about zero current. Sidebands around 1 kHz are present, however, at frequencies of 750, 850, 1050 and 1150 Hz. Up to 700 Hz the harmonics are well suppressed, above 1 kHz the odd harmonics extend at more or less constant amplitude up to very high frequency (depending on the switching rise and fall times).

The RC snubber across the input supply is necessary to avoid a large voltage spike as the switches change their configuration between successive half-cycles. The current in the switches that are opening falls more rapidly than it rises in the switches closing, and the excess current is diverted to the capacitor.

## Summary

In this chapter the basic types of single-phase inverter have been considered. Inverters may have either a voltage or a current source, depending on whether the load is to be supplied with an alternating voltage or an alternating current source. Care must be taken with the nature of the load when using switching inverters, since the rapid changes of voltage or current may produce large current or voltage spikes. If the load is a source of reactive power then the inverter must be able to absorb this power, which for a voltage-source inverter means current flowing back into the voltage source while for a current-source inverter it implies a reversed source voltage. The load voltage (or current) and the harmonic content of the output of a switching inverter may be controlled by using pulse-width modulation. Filtering is, however, still necessary to remove components around and above the switching frequency, but the filter components (capacitors and inductors) may be much lower value than would be required to filter low-order harmonics of the line frequency.

## Self-assessment questions

10.1　A single-phase half-bridge inverter has source voltages of $+100$ V and $-100$ V. The inverter operates with a square voltage waveform at a frequency of 50 Hz. The load is a resistance of 20 $\Omega$ in series with an inductance of 100 mH. What will be the peak current?

10.2　The inverter described in Question 10.1 is operated with bipolar sinusoidal pulse modulation at a frequency of about 20 kHz. The maximum duty ratio at the peak of the sine wave is 0.9. Neglecting losses in the switches, estimate the r.m.s. amplitude of the fundamental frequency component of the output voltage and the output current. Is it reasonable to assume that if the load is as in Question 10.1, the current at the switching frequency may be neglected, and hence the fundamental component of the load current is a good approximation to the total current?

10.3　A single-phase full-bridge inverter is single-pulse modulated. The operating frequency is 50 Hz, the source potential is 200 V, and each pulse has a duration of 6 ms. What is (a) the r.m.s. load voltage and (b) the r.m.s. magnitude of the fundamental?

10.4　A single-phase full-bridge inverter with a source voltage $V_S$ is unipolar multiple-pulse modulated with a constant duty ratio $D$. Find the r.m.s. magnitude of the load voltage.

10.5   A single-phase current-source inverter produces a square wave current at a frequency of 400 Hz. The source current is 5 A and the load comprises a 10 Ω resistor in parallel with a 100 $\mu$F capacitor. What is the peak load voltage?

10.6   For the inverter and load described in Question 10.5, sketch the load voltage and source voltage, showing the times at which the voltages have their maximum, minimum and zero values.

10.7   A single-phase full-bridge single pulse inverter operates with a source voltage of 150 V. The load voltage is to have a fundamental component with an r.m.s. value of 110 V at a frequency of 60 Hz. What is the appropriate value for the time between pulses?

10.8   For the inverter of Question 10.7, find the r.m.s. amplitude of the third and fifth harmonics.

## Tutorial questions

**10.1**   For a voltage-source single-phase full-bridge inverter explain the role of the reverse diode across the switches.

**10.2**   Why should a voltage-source inverter not be used with a highly capacitive load?

**10.3**   What is the switching sequence for a single-phase voltage-source full-bridge inverter if the load voltage is to be single-pulse modulated?

**10.4**   What is the switching sequence for a single-phase current-source full-bridge inverter if the load current is to be single-pulse modulated?

**10.5**   Why can a load commutated current-source inverter not produce a load current in phase with the load voltage?

**10.6**   Why would multiple-pulse modulation be preferred to single-pulse modulation if a wide range of control is required over load voltage (or current)?

**10.7**   For a single-phase full-bridge inverter, why would unipolar PWM normally be preferred to bipolar PWM?

**10.8**   What is an appropriate sequence for the operation of the switches in a single-phase full-bridge inverter when it is using unipolar modulation to produce (a) a positive voltage and (b) a negative voltage?

## References

Mohan, N., Undeland, T. M. and Robbins, W. P., *Power Electronics: Converters, Applications and Design*, John Wiley, New York, 1995.

Rashid, M. H., *Power Electronics: Circuits, Devices and Applications*, 2nd edition, Prentice Hall, Englewood Cliffs, NJ, 1993.

Vithayathil, J., *Power Electronics: Principles and Applications*, McGraw-Hill, New York, 1995.

## CHAPTER 11

# Drives for small electric motors

## 11.1 Introduction

One of the major areas of application for power electronics is the control of electric motors. At the high-power end are very large motors for industrial and traction applications (with power measured in megawatts) down to small electric motors in consumer electronic equipment, e.g. drive motors in tape recorders, where the power may be less than a watt. Across this range there is a requirement for accurate control of speed, torque, and in some cases, the position of the armature. There are many types of electric motor in use, but probably the most widespread are d.c. motors, brushless d.c. motors, induction motors and synchronous motors. These normally produce rotational motion, although for some purposes motors producing linear motion are required.

Here only drives for low-power motors will be considered, although the techniques required for larger power motors differ mostly in the types of electronic switch required to handle the power. The main areas of application for small electric motors, where electronic control is required, are in servo-systems for controlling mechanical systems, electronic equipment, such as video-tape recorders or computer disk drives, and in many small items of domestic or light industrial equipment, e.g. washing machines or air-conditioning units.

## 11.2 D.C. motors and their drives

### 11.2.1 Small d.c. motors

A d.c. electric motor has three essential components: the armature or rotor, the commutator and the stator. The armature is wound with several coils of wire or windings which carry electric currents. These currents interact with the magnetic field produced by the stator and produce the torque which drives the motor. The commutator is required to transmit the electric current to the windings of the armature as it rotates, and to switch the direction of the current in the windings to ensure a unidirectional torque and continuous rotation.

D.C. motors are divided into two main groups, those in which the magnetic field in the stator is produced by electric current in a *field winding* and those in which it is produced by a permanent magnet. For large power d.c. motors a field winding is

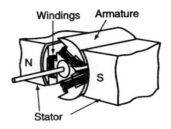

**Figure 11.1** *A simple d.c. motor*

essential to produce a magnetic field of sufficient strength and volume. Motors with a power of up to about 1 kW may use permanent magnets. A permanent-magnet motor is generally smaller, cheaper and easier to use, but the use of field windings gives greater power and much greater flexibility. Here only permanent-magnet motors will be considered.

There is a range of possible structures for d.c. motors, but the simplest is illustrated in Figure 11.1. The stator provides two magnetic poles between which the armature rotates. The armature comprises a cylindrical soft iron core, laminated to reduce eddy current losses, in which slots are cut which carry the windings. To ensure that the motor can start in any position at least three slots (and windings) are required. In practice, there may be many more slots and windings to produce a more constant torque as the motor rotates.

A simple model for the electrical side of a permanent-magnet d.c. motor is illustrated in Figure 11.2. Neglecting the complications associated with the discrete changes which occur during commutation, the electrical circuit of the armature can be considered to consist of the resistance of the windings in series with their inductance and the *back-e.m.f.* produced by the rotation of the windings in the magnetic field. The back e.m.f. is related to the speed of rotation by the constant $K_E$, and is given by

$$E = K_E \Omega_R \tag{11.1}$$

where $\Omega_R$ is the angular speed of the motor shaft.

On the mechanical side, the motor produces a torque, $T$, proportional to the current in the windings of the armature, i.e.

$$T = K_T I_A \tag{11.2}$$

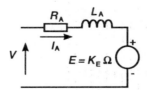

**Figure 11.2** *A circuit model for a d.c. motor*

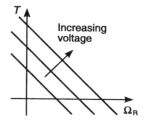

**Figure 11.3** *The torque against speed characteristics of a d.c. motor*

The mechanical power produced by the motor is $T \, \Omega_R$, while the electrical power producing the mechanical power is the armature current, $I_A$, multiplied by the back e.m.f. $E$. Hence from equations (11.1) and (11.2)

$$K_E = E_T$$

provided, of course, both are quoted in consistent units. For an electric motor with field windings the value of $K_E$ and $K_T$ will be proportional to the flux produced by the field windings.

If the voltage across the terminals of the motor is $V$, then in the steady state

$$V = R_A I_A + E \tag{11.3}$$

where $R_A$ is the resistance of the armature. Combining equations (11.1) and (11.2) with (11.3) gives

$$T = \frac{V K_T - K_E K_T \Omega}{R_A} \tag{11.4}$$

The variation of torque with speed, as given by equation (11.4), is plotted in Figure 11.3. A permanent-magnet d.c. motor has a large torque at low speed, a maximum speed that is controlled by the applied voltage and a direction of rotation controlled by the polarity of the applied voltage. These characteristics make this type of motor very suitable for servo-systems.

### 11.2.2 Driving permanent-magnet d.c. motors

For use in a servo-system a permanent-magnet d.c. motor should be able to rotate in either direction with the speed and torque under control. This requires a variable voltage supply whose polarity may be reversed. The motor may also operate as a brake if the direction of the armature current is reversed. The four modes of operation are illustrated in Figure 11.4. When the direction of rotation and the sign of the current are either both positive or both negative the motor supplies power to the load. With positive rotation and negative current or negative rotation and positive current, the motor acts as a brake. Thus a controller must be able to source or sink current of either polarity. To control the speed it must control the voltage, while to control the torque it must control the current.

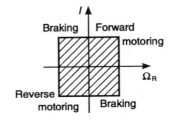

Figure 11.4   *The operating modes for a d.c. motor*

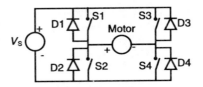

Figure 11.5   *A bridge driver circuit for a d.c. motor*

Control using a d.c. linear amplifier with a split power supply, similar to the audio amplifiers to be discussed in Chapter 12, is possible but is not usual. A linear amplifier might be considered for a very low power motor but the efficiency would be low and the benefits small. The more usual solution is to use pulse-width modulation, usually with a full-bridge configuration (Figure 11.5) to allow bidirectional operation. The switches would be bipolar transistors or MOSFETs for low-power motors. IGBTs or GTOs might be used at much higher power.

The bridge may be operated to give either a unipolar or a bipolar pulse-width modulation in the way described for bridge inverters. To drive the motor in the positive direction S4 would be turned on then the motor voltage, pulse modulated by switching S1 and S2 alternately to set the voltage high or low. This is represented by the simplified circuit in Figure 11.6. Current may flow in either direction. If the motor is not at rest and hence is generating an e.m.f. the armature current will be continuous. Neglecting the voltage drop across the switches, the mean voltage across the motor is given by $DV_S$, where $D$ is the duty ratio of the PWM.

The alternative switching scheme is a bipolar drive, with Q1 and Q4 switching alternately with Q2 and Q3. The voltage applied to the motor switches between $+V_S$ and $-V_S$ with a duty ratio $D$. Current may flow in either direction, depending on the

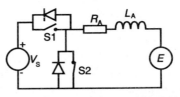

Figure 11.6   *A simplified circuit for unipolar pulse-width modulated drive in the forward direction*

mean applied voltage and the back-e.m.f. The current will be continuous, even if the motor is stationary. The mean load voltage across the motor is given by

$$\bar{V} = (2D - 1)V_S \tag{11.5}$$

If the pulse modulation frequency is sufficiently high, the inertia of the armature will ensure that its rotational speed will be almost free of ripple. In this case the back-e.m.f, $E$, may be considered to be a constant. If the motor voltage is $v(t)$ then the armature current, $i_A(t)$, is given by

$$v(t) = E + R_A i_A(t) + L_A \frac{di_A(t)}{dt} \tag{11.6}$$

Since the armature current is continuous, the mean load current will be given by

$$\bar{I}_A = \frac{\bar{V} - E}{R_A} \tag{11.7}$$

The magnitude of the ripple current will depend on the inductance of the armature, the switching frequency and the duty ratio. For unipolar modulation, assuming that the armature resistance, $R_A$, is small and that the ripple current is limited by the inductance of the armature, the peak-to-peak ripple current, found by integrating equation (11.6), is given by

$$\Delta i = \frac{V_S(1 - D)DT}{L_A} \tag{11.8}$$

where $T$ is the switching period. For bipolar modulation the peak-to-peak ripple current will be twice this value. The modulation frequency should be such that it is high enough to avoid significant modulation of the rotational speed and excessive ripple current, with its associated power dissipation. A value just above the audio range, 20–25 kHz, will avoid excessive audio noise.

A range of integrated circuits are available to assist with the construction of controllers for small d.c. motors. Full- and half-bridge drivers using either bipolar or DMOS switches are available to control currents of 1–2 A at a voltage of about 40 V, with control provided by 5 V logic. Complete controllers for pulse-width modulation control of small motors are also available, which provide for closed-loop control of the motor current using pulse-width modulation, together with the power drivers, in a single package.

## 11.3 Brushless d.c. motors

Permanent-magnet, d.c. motors have characteristics that are well suited to variable-speed drives or servomotors. Their main disadvantage is that the brushes used for the mechanical commutation are inevitably subject to wear and require maintenance. The brushless d.c. motor avoids this difficulty by using solid-state

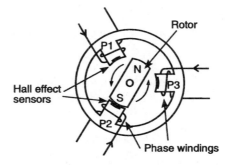

**Figure 11.7** *A simplified schematic diagram of a three-phase brushless d.c. motor*

switches to perform the commutation. In the conventional permanent-magnet d.c. motor the field magnet is part of the stator and the rotor carries the windings in which the current must be commutated. In the brushless d.c. motor the arrangement is reversed. The permanent field magnet is part of the rotor and the windings are part of the stator. This avoids the need for brushes to carry the current to rotating windings.

The basic arrangement of a three-phase two-pole brushless d.c. motor is shown in Figure 11.7. To provide continuous rotation the three windings must be energized in sequence. It is very closely related to a synchronous three-phase a.c. motor, but the difference is that in a brushless d.c. motor the position of the rotor is sensed, usually using Hall effect sensors, to detect the position of the poles of the rotor, although optical sensors may be used instead.

To cause the rotor to rotate anti-clockwise from the position shown in Figure 11.7 the pole P1 should be energized to give a south pole while P3 should give a north pole. Pole P2 should be neutral. When the rotor has rotated through 60°, pole P1 should be de-energized, and pole P2 energized as a south pole, to continue the rotation. This sequence is continued to produce continuous rotation, and the required currents in each of the three-phase windings are shown in Figure 11.8.

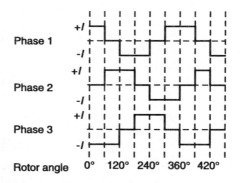

**Figure 11.8** *The sequence of phase currents for a brushless d.c. motor*

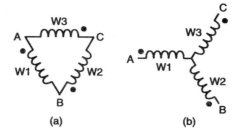

**Figure 11.9** *Connections for the phase windings of a three-phase brushless d.c. motor (a) Δ connected and (b) Y connected*

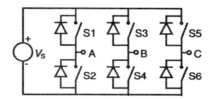

**Figure 11.10** *A three-phase bridge inverter*

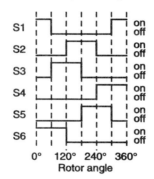

**Figure 11.11** *The switching sequence for driving a three-phase brushless d.c. motor in an anti-clockwise direction*

The windings may be either Δ or Y connected, as shown in Figure 11.9, where the 'dots' show the top end of each of the windings. The three-phase connections will be energized from a three-phase inverter bridge as shown in Figure 11.10. For a motor with Y-connected windings the sequence for operating each of the switches is shown in Figure 11.11 for anti-clockwise rotation. With this switching sequence, Y-connected windings are energized for 120° per half-cycle.

In order to reverse the direction of rotation the direction of the currents in each winding must be reversed by swapping the switching sequence of the upper and lower switches in each 'leg' of the inverter (i.e. swap S1 and S2, S3 and S4, and S5 and S6). The direction of phase rotation is reversed, since the motor now rotates in the opposite direction and the Hall switches operate in the reverse order. To control the speed, pulse-width modulation may be used as for a conventional d.c. motor.

The logic and control functions necessary to control brushless d.c. motors are quite complex, but several manufacturers produce integrated circuits to perform the necessary logic functions to decode the outputs from the rotor position sensors and to provide pulse-width modulation. The PWM is frequently achieved by sensing the current in the windings, and cutting it off, for a set period, when it exceeds a value determined by the control voltage. The inductance of the windings determines how quickly the current rises, and hence the frequency. This method of generating pulse-width modulation controls the motor current, rather than the voltage, which in addition to its simplicity, has the advantage that the current is limited, protecting the motor and the switches.

Only a simple three-phase two-pole motor with bipolar drive has been described here. Many other configurations exist. The rotor may have more than two magnetic poles, so that the mechanical rotation per cycle of the switched current is only a fraction of a complete turn. A four-pole motor will rotate by half a turn per cycle of the electrical drive. This gives smoother rotation. The drive may be unipolar rather than bipolar, which will simplify the drive logic, or there may be two or four phases. Obviously, the logic for decoding the signals from the position sensors, and the power switching circuit itself, have to be matched to the type of motor and the location of the sensors.

## 11.4  Variable-speed drives for induction motors

### 11.4.1  Three-phase induction motors

Three-phase induction motors using squirrel-cage rotors are widely used because they are cheap and robust. The windings of the motor are in slots on the inside of the stator, which is tubular in shape, and are arranged to produce a rotating magnetic field in the air gap between the stator and the cylindrical rotor. The rotor is made from disk-shaped laminations stacked to form a cylinder. Round the periphery of the rotor a number of slots are cut, and in these slots lie electrically conducting bars which are connected together electrically at the ends (the squirrel cage).

In operation the currents in the three-phase windings induce a magnetic field of constant flux density which rotates at the synchronous speed $\Omega_s$ rad s$^{-1}$. For a $p$-pole motor the synchronous speed is given by

$$\Omega_S = \frac{2\omega}{p} \tag{11.9}$$

where $\omega$ is the frequency of the applied a.c. The rotor rotates at a frequency $\Omega_R$, slightly lower than the synchronous frequency, and the difference $\Omega_{sl}$ is the slip speed:

$$\Omega_S = \Omega_R + \Omega_{sl} \tag{11.10}$$

The slip introduces motion between the rotating flux and the conductor bars on the rotor, and hence induces circulating currents which are proportional to the rate of change of flux or the slip speed. These circulating currents produce a field that rotates at the slip speed relative to the rotor, and hence at the synchronous speed relative to the stator. The interaction between the flux due to the rotor currents and the flux due to the currents in the stator windings produces the torque which drives the motor. For small values of slip the torque is proportional to the slip speed. Thus with no load the rotor speed approaches the synchronous speed, and decreases almost linearly with torque up to the rated torque for the motor. Induction motors are usually designed so that at the rated frequency and voltage the slip speed at the rated torque is small compared with the synchronous speed. $\Omega_{sl}$ is typically of order 5% of $\Omega_S$.

### 11.4.2  Variable-speed drives for three-phase induction motors

To control the speed of an induction motor it is necessary to vary the frequency of the three-phase excitation. However, as the frequency is reduced, the magnetization current increases, increasing the flux in the air gap. To avoid saturation of the core the applied voltage must be reduced as the frequency is reduced. Induction motors are normally designed to operate at the power-line frequency (50 or 60 Hz) and at a rated voltage and torque. If such a motor is to be used with a variable-speed drive it makes good sense to keep the flux constant at its value under rated conditions. To achieve this, the ratio of the applied voltage, $V$, to the frequency, $\omega$, should be kept constant, so that the voltage is given by

$$V = \frac{V_R \omega}{\omega_R} \qquad (11.11)$$

where $\omega_R$ is the rated frequency and $V_R$ is the rated voltage.

Induction motors are optimized for operation at their rated frequency, voltage and torque, and are usually designed so that under these conditions the slip speed is small, typically only a few per cent of the synchronous speed. If the motor is run at reduced frequency and voltage, the same torque is available with the same slip speed. If the frequency is significantly below the rated frequency, then the slip speed will be a much greater fraction of the synchronous speed.

Since induction motors are usually designed to operate from the a.c. power line they are designed for sinusoidal excitation. The inverter necessary for variable-speed operation of an induction motor must therefore produce a sinusoidal, variable voltage three-phase output. The voltage-source inverter of Figure 11.10 can be used with sinusoidal pulse-width modulation to generate a suitable variable frequency and amplitude drive.

In Chapter 10 the operation of a single-phase sinusoidal pulse-width modulated inverter was described. In that case the switches in the two legs of the bridge were controlled from two pulse-width modulators using reference sine waves in anti-phase. The same technique can be used with a three-phase inverter, but with three pulse-width modulators and with three sine wave references, delayed in phase by 0°, 120° and 240°. The frequency and amplitude of the output may be controlled by controlling the amplitude and frequency of the reference signals.

The control of a.c. induction motors is rather complex, and is probably an example where there are very significant advantages to using a microcontroller to generate the switch control signals. However, despite the relative complexity of the drive, the very substantial advantages of induction motors, in terms of their cost and robust nature, have ensured that variable-speed drives for induction motors have become well established.

## 11.5  Synchronous a.c. motors

A synchronous a.c. motor differs in essence from a brushless d.c. motor only in the way power is supplied to the phase windings on the stator. For a brushless d.c. motor, the alternating drive to the windings is produced by switching the current synchronously with the position of the rotor. For a synchronous a.c. motor, a.c. is supplied to the windings with an appropriate phase relationship between the phases, and, provided the motor is not stalled, the rotor will rotate at a frequency determined by the applied a.c. The stator itself may have a field produced by a permanent magnet or in large sizes by a field winding on the rotor.

The rotation of the rotor generates a back-e.m.f. in the phase windings. For a motor designed for use with sinusoidal a.c. the phase windings are distributed between a number of slots in such a way that the back-e.m.f. at constant rotational velocity is sinusoidal. This will ensure that the current drawn by the motor is also sinusoidal, and give a good power factor. Such motors should be operated only with a sinusoidal waveform, as described in Section 11.4.2 for an induction motor. This will avoid excessive power dissipation due to a poor harmonic factor. Synchronous motors may be designed for use with a square waveform, but this requires a different winding arrangement for the rotor.

To generate a three-phase square waveform the same three-phase inverter configuration is used (Figure 11.10), but with a simpler switching sequence than for a sinusoidal pulse-width modulation. Each leg of the inverter is switched to the source voltage for half the cycle and to zero for the other half-cycle. Figure 11.12 shows the voltages between the three phases and the supply negative line, and the voltage between phases A an B. The inter-phase voltages show a stepped waveform.

## 11.6  Stepper motors

### 11.6.1  Variable-reluctance stepper motors

Stepper motors differ from conventional motors in that they move not continuously but in steps between well-defined positions. The two most common types of stepper

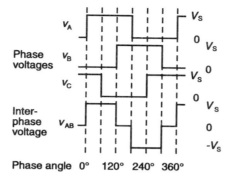

**Figure 11.12** *The voltages between phases A, B and C and the supply negative line for a three-phase square wave inverter, and the inter-phase voltage V_AB*

motor are the permanent-magnet stepper motor, which is the type most often used for small stepper motors, and the variable-reluctance stepper motor, which can be scaled up to much larger sizes. Both are, in principle, types of synchronous motor. They differ from ordinary synchronous motors in the details of the internal design of the magnetic circuit, which is designed to provide motion between fixed positions, and in the way they are driven.

A simplified diagram of a variable-reluctance (VR) motor is shown in Figure 11.13. There are three phases with six teeth on the stator and four teeth on the rotor. Phase A is energized, and the magnetic field holds the rotor teeth labelled 1 in line with the stator teeth of phase A to minimize the energy stored in the magnetic field. To rotate the rotor anti-clockwise, phase B is energized and phase A is turned off. The rotor turns through 30° to align the teeth labelled 2 with the poles of phase B. To make the next step, phase C is energized and phase B de-energized, rotating the rotor by a further 30°. To reverse the direction the phase rotation is reversed. The sequences for the phase currents to produce anti-clockwise and clockwise rotation are shown in Figure 11.14. VR stepper motors require only a unipolar supply and the current in the phase windings does not need to be reversed.

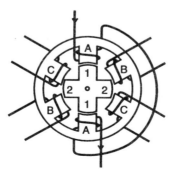

**Figure 11.13** *A simple variable-reluctance stepper motor*

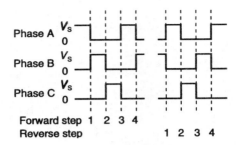

Forward step 1  2  3  4
Reverse step              1  2  3  4

**Figure 11.14**  *The current in the three-phase windings of a variable-reluctance stepper motor to produce anti-clockwise and clockwise rotation*

The construction of practical VR stepper motors is usually rather more complex than shown in Figure 11.13. To reduce the step size many more teeth are required, and a separate rotor and stator (referred to as a stack) may be used for each phase. At least three phases are needed to determine the direction of rotation. More could be used but are not necessary.

### 11.6.2  Permanent-magnet stepper motors

Permanent-magnet (PM) stepper motors use quite a complex structure to achieve a small step size, again using a rotor and stator with a large number of teeth. However, the operation can be understood in terms of the simplified model shown in Figure 11.15. The stator has two phases, which are driven with a bipolar currents. If the current in phases A and B is such that both the poles marked a and b are south poles, while a' and b' are north poles the rotor will align as shown. If the polarity of a is reversed, then the rotor will rotate through 90° in an anticlockwise direction. This completes a *full step*. Reversing phase B will continue the rotation by a further step.

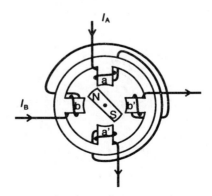

**Figure 11.15**  *A simple, permanent-magnet stepper motor*

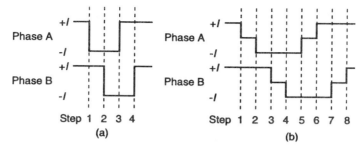

**Figure 11.16** *The phase currents for a two-phase permanent-magnet stepper motor: (a) for full stepping; (b) for half stepping*

It is also possible to produce *half-steps*. Starting from the situation of Figure 11.15, if phase A is turned off the rotor will move anti-clockwise by 45° for one half-step. Energising phase A in the reverse direction will provide the second half step. The current sequences for rotation in the anticlockwise direction with full and half-steps is shown in Figure 11.16. To reverse the direction the phase rotation must be reversed.

### 11.6.3 Driving stepper motors

Variable-reluctance stepper motors require a sequence of unipolar currents in the phase windings. When the motor is stationary the current in the phase winding that is energized is limited only by the series resistance and the applied voltage. To produce a rapid step from one position to the next it is important that the current rises rapidly in one phase and decreases rapidly in the other. Thus it is desirable that the inductance is low. Simple transistor switches can be used to switch the unipolar currents, as shown in Figure 11.17. The phase windings are $L_A$, $L_B$ and $L_C$. The current in the windings is limited by their resistance and the source voltage $V_S$. If necessary, extra resistance may be added. When a switch turns off the current is commutated to the freewheeling diode. The resistances, $R$, may be added to reduce the time constant for the decay of the current and speed up the stepping. Pulse-

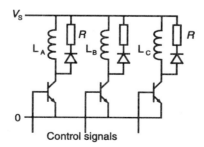

**Figure 11.17** *A drive circuit for a three-phase variable-reluctance stepper motor*

width modulation may be used to limit the current in the phase windings, without wasting power, by adding a series resistance. Pulse-width modulation has the added advantage that the switch may be left on continuously while the current increases to its operational value, allowing an increased voltage across the windings during this part of the cycle and giving a more rapid response.

Small stepper motors are generally of the permanent-magnet or hybrid variety. The hybrid stepper motor has a permanent-magnet rotor, but its operation relies on both the interaction of the magnetic rotor with the field produced by the stator and the forces due to the change in reluctance of the magnetic circuit. The drive requirements are essentially the same for both types. These types of stepper motor are used extensively in computer peripherals such as printers and disk drives.

The permanent-magnet stepper motor requires a bipolar drive, which implies either a split supply and a half-bridge driver or, more usually, a full-bridge driver. Small sizes of stepper motor often overcome this complication by the use of *bifilar windings*. Each phase is wound with two windings, with the windings connected together to give a centre-tapped winding. This permits a bipolar magnetic field to be produced with a unipolar drive, using the circuit shown in Figure 11.8. Either polarity of field may be produced by switching on one of the two transistors associated with each phase. For half-stepping, to give zero current neither transistor is turned on. Frequently the current limiting resistors R are the winding resistances or a resistance internal to the motor. This type of stepper motor with bifilar windings is frequently described as a four-phase unipolar motor.

Several manufacturers produce integrated circuits for driving permanent-magnet stepper motors. These contain a state-machine to produce the switching sequence necessary to drive the motor, so that the user needs only to provide a logic level to determine the direction and a clock pulse to induce the motor to step. These will generally drive a small bifilar wound stepper motor directly. The more sophisticated ICs will permit the selection of half or full stepping. If more power is needed, or a bipolar drive or PWM control of the current, then there are ICs available to provide these facilities.

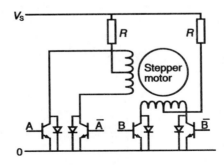

**Figure 11.18**  *A drive circuit for a four-phase unipolar stepper motor*

# Summary

This chapter has attempted to provide a brief overview of the use of power electronics to control electric motors. While d.c. motors have traditionally been widely used in variable-speed and servo-systems, the flexible control of frequency and voltage that is possible using electronic switching and pulse-width modulation has made it possible to use a.c. motors in circumstances where previously d.c. motors would have been required. Here attention has been concentrated on low-power drives, such as might be used in disk drives, video recorders and similar equipment. Power electronics has, however, made a similar impact on high-power drives for traction and heavy industrial applications.

## Self-assessment questions

11.1 A small, permanent-magnet d.c. motor has a torque constant of 0.03 N m $A^{-1}$ and an armature resistance of 3 $\Omega$. If the motor is to rotate at 2000 r.p.m., with a load of 0.04 N m, what will be the motor current and terminal voltage?

11.2 The motor of Question 11.1 is driven using pulse-width modulation from a voltage source of 15 V. What duty ratio will give a rotational speed of 2000 r.p.m. if the modulation is (a) unipolar and (b) bipolar?

11.3 A motor with an armature resistance of 5 $\Omega$ and an inductance of 5 mH is pulse-width modulated at a frequency of 20 kHz. What will be the maximum peak-to-peak amplitude of the ripple current if the modulation is unipolar and the source voltage 20 V?

11.4 A stepper motor for bipolar drive has an inductance of 8 mH and a winding resistance of 1.2 $\Omega$. The rated phase current is 3.5 A. A series resistance is used to limit the current to this value. What will be the required series resistance and the time constant for the rise of current if the motor is used with (a) a 6 V supply and (b) a 24 V supply? In each case what is the power dissipated in the resistor?

11.5 The stepper motor described in Question 11.4 is operated from a supply of 24 V and uses PWM to limit the current to 3.5 A. The modulator starts to operate only when the current teaches this limit. How long will it take to reverse the current in a phase winding, from $-3.5$ A to 3.5 A, when the phase polarity is reversed?

## Tutorial questions

**11.1** Explain why for a d.c. motor the torque is proportional to the current while the speed is proportional to the voltage.

**11.2** How would the switches of the bridge circuit of Figure 11.5 be configured to give drive in the reverse direction and braking when the motor is rotating in the reverse direction?

**11.3**  With reference to Figure 11.6, why are two switches used in the pulse-width modulator when it is used to control the motor speed? What would be the consequence of leaving S2 open all the time?

**11.4**  How could the three-phase brushless d.c. motor described be operated using only a unipolar current drive?

**11.5**  Figure 11.8 shows the sequence of the phase currents for anti-clockwise rotation of a three-phase brushless d.c. motor. What would be the sequence for clockwise rotation, starting from the same position of the rotor?

**11.6**  The sequence in which the switches must be operated for a three-phase Y-connected, brushless d.c. motor is shown in Figure 11.11. How would the sequence have to be modified if the phases were Δ connected?

**11.7**  Will the polarity of the current have any effect on a variable-reluctance stepper motor?

**11.8**  A bifilar wound stepper motor has the connections for the two windings for each phase brought out separately. If the motor is used with these two windings either in series or in parallel, for each phase, can a greater torque be produced than with the usual unipolar drive, assuming that the current is limited by the power dissipation?

**11.9**  Explain how pulse-width modulation may help to achieve the highest step rate from a stepper motor.

## References

Kenjo, T., *Stepping Motors and their Microprocessor Controls*, Clarendon Press, Oxford, 1984.

Kenjo, T. and Nagamori, S., *Permanent-Magnet and Brushless DC Motors*, Clarendon Press, Oxford, 1985.

Krause, P. C. and Wasynczuk, O., *Electromechanical Motion Devices*, McGraw-Hill, New York, 1989.

Mohan, N., Undeland, T. M. and Robbins, W. P., *Power Electronics: Converters, Applications and Design*, John Wiley, New York, 1995.

Rashid, M. H., *Power Electronics: Circuits, Devices and Applications*, 2nd edition, Prentice Hall, Englewood Cliffs, NJ, 1993.

## CHAPTER 12

# Audio-frequency power amplifiers

## 12.1 Introduction

Audio-frequency power amplifiers are required for the reproduction of recorded music, amplifying electronic instruments and public address systems, and amplifiers for these applications will be considered in this chapter. There are also a few more specialized and unusual applications in the areas of control and actuation, where magnetic or electrostatic actuation devices must be driven at frequencies from a few Hz up to many kHz. For the usual type of audio application, the power that must be delivered to the load may vary from a few watts to several kilowatts, with a flat frequency response from about 20 Hz to 20 kHz. Unlike most power electronic systems, the most critical design consideration is not power efficiency but low distortion. For this reason, switching techniques using pulse-width modulation have made little impact, and traditional linear amplifier designs are the most popular. In this chapter consideration will be given to the audio amplifiers that use the traditional class A and class B output stages. Class D amplifiers that use pulse width modulation will also be considered, to show their potential advantages and the practical difficulties that have limited their use.

## 12.2 Push–pull output stages

Practical power output stages for linear amplifiers generally use a push–pull topology as illustrated in Figure 12.1. The circuit uses one transistor to pull up the load voltage, and to source current, while the other transistor is used to pull down the load voltage, and to sink current. The most common configuration is based on the simplified circuit of Figure 12.1, which is a complementary emitter follower circuit. The transistors may be either bipolar transistors as shown, or MOSFETs, although here it will be assumed that bipolar transistors are used. The question of whether bipolar transistors or MOSFETs are more suitable for audio amplifier applications will be considered later. The load current is the difference between the emitter currents in Q1 and Q2. Since the circuit is basically an emitter follower, the output impedance is low. This complementary emitter follower stage may operate in either class A or class B. In a class A output stage the quiescent current is set so that for the full range of output voltage swing the current does not fall to zero in either transistor.

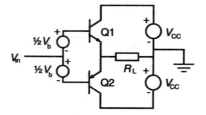

**Figure 12.1** *The basic complementary emitter follower*

For a class B stage the quiescent current is set to a low value, and except for very small signals one transistor will supply current to the load while the other is cut off.

### 12.2.1 Class A operation of a push–pull stage

For class A operation the current must not fall to zero in either transistor for any value of the load voltage. The bias voltage, $V_b$, is adjusted to set the quiescent current, so that over the whole range of output voltage, and with the minimum load impedance, this condition is met. The current in the load resistor is given by

$$I_L = -I_{E1} - I_{E2} \tag{12.1}$$

where $I_{E1}$ and $I_{E2}$ are the emitter currents of Q1 and Q2, which, by convention, are taken to flow into the emitter, hence the $I_{E1}$ is negative and $I_{E2}$ positive. For a small change, $v_{io}$, between the input node and the output node the small change in output current is given by

$$i_L = (g_{m1} + g_{m2})v_{io} \tag{12.2}$$

where $g_{m1}$ and $g_{m2}$, are the transconductances of the transistors Q1 and Q2 at their respective collector currents and the differences in magnitude between the base and

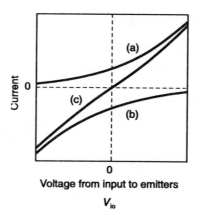

**Figure 12.2** *The transfer characteristic of two transistors in class A, push–pull circuit. (a) The upper transistor; (b) the lower transistor; (c) the resultant of the two transistors*

emitter currents have been neglected. The effective transconductance, $G_m$, of the two transistors is therefore the sum of their transconductances, $G_m = g_{m1} + g_{m2}$. The load voltage is given by

$$v_{out} = R_L G_m v_{io} = R_L G_m (v_{in} - v_{out})$$

hence,

$$A_v = \frac{v_{out}}{v_{in}} = \frac{G_m R_L}{1 + G_m R_L} \tag{12.3}$$

The transconductance of the two transistors varies with the current in the transistors, but as one increases, the other decreases, reducing the variation in $G_m$ and improving the linearity. This is illustrated in Figure 12.2, where the emitter currents of each transistor are shown schematically together with their sum. The transconductance of the pair, Gm, is obtained from the slope of the sum of the emitter currents.

### 12.2.2 Class B operation of a push–pull stage

The class A output stage is capable of good linearity but is very inefficient, because the quiescent current must exceed the peak load current. The class B stage uses the same circuit topology (Figure 12.1) but with a very small quiescent current. Transistor Q1 turns on to supply positive load current and Q2 turns on to supply negative current. Only one transistor conducts significantly at one time. To achieve a smooth transition from Q1 conducting to Q2 conducting and avoid the *cross-over distortion* requires careful adjustment of the bias voltage. If the bias voltage is too small, one transistor will cease conduction before the other starts to conduct. If it is too large then there will also be a non-linearity due to both transistors contributing to the effective transconductance, as in the class A stage.

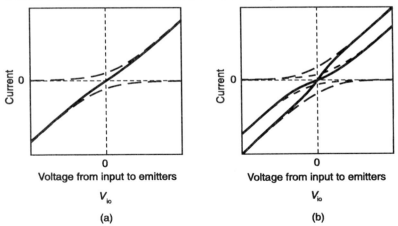

**Figure 12.3** *The transfer characteristics of two transistors in a class B push–pull circuit. (a) The optimum quiescent current; (b) a quiescent current that is too large or too small*

The transfer of current from one transistor to the other in class B is illustrated in Figure 12.3, which shows schematically the current in each transistor together with the resultant load current, as a function of the voltage between the input and the emitters, $V_{io}$. For the correct value of bias, the transfer function may be reasonably linear, as illustrated in Figure 12.3(a). If the bias is too high, or too low, the transfer function will show an increase or decrease in the gradient near the origin, i.e. an increase or decrease in $G_m$. The optimum value of the bias depends on the transistor characteristics, and especially on the resistance in series with the emitter. As practical circuits always include extra emitter resistance to improve the bias stability, the optimum bias value depends on the circuit. Here, this optimum bias condition is referred to as class B, although this mode of operation is sometimes described as class AB, since for small signals both transistors conduct.

## 12.3  Power dissipation in push–pull output stages

### 12.3.1  Power dissipation in class A amplifiers

The power dissipation will be analyzed for a very simple amplifier assuming the circuit of Figure 12.1. Dual power rails of $\pm V_{CC}$ are assumed, and the load is assumed to be a pure resistance, $R_L$. It is also assumed that the load voltage can swing up to the value of power rails but this is rather simplistic. The factor that usually limits the voltage swing is not the saturation voltage of the output transistors, which may well be less than 1 V, but the maximum voltage swing at the base of the transistors, since this is usually derived from the same power rails.

For a class A amplifier it is necessary to select the quiescent current of the transistors so that the current does not fall to zero in either transistor for the lowest value of load and maximum voltage swing. Neglecting the difference between collector current and emitter current, the maximum voltage swing, $V_{out}$, for a quiescent collector current, $I_Q$, is given by

$$V_{out} = 2I_Q R_L \tag{12.4}$$

assuming that the transistors do not saturate. It follows from equation (12.4), that the minimum value of the quiescent current for the amplifier to stay in class A, with a load resistance $R_L$, and if the load voltage swings to the power rails, is given by

$$I_Q = \frac{V_{CC}}{2R_L} \tag{12.5}$$

For a class A amplifier the average value of the current drawn from each supply is equal to the quiescent current, assuming there is no d.c. current to the load. If the quiescent current has the minimum value, given by equation (12.5), and given that the supply rail voltage is constant, the power drawn by the class A output stage is given by

$$P_{DC} = 2V_{CC}I_Q = \frac{V_{CC}^2}{R_L} \tag{12.6}$$

The power drawn does not depend on the signal level.

The power supplied to the load will depend on the level and the waveform of the signal. Assuming a sinusoidal signal of amplitude, $V_{out}$, the power delivered to the load is given by

$$P_{out} = \frac{V_{out}^2}{2R_L} \tag{12.7}$$

If the maximum amplitude is $V_{CC}$, from equations (12.6) and (12.7), the maximum efficiency, $\eta$, of the push–pull class A stage is

$$\eta = \frac{P_{out}}{P_{DC}} = 0.5$$

In a real output stage the efficiency will be somewhat less than 0.5 because of the more limited voltage swing and power lost in the emitter resistors that are essential to ensure the stability of the quiescent current.

An interesting point to note, in passing, is that the power dissipation in the power transistors is greater when there is no signal power, since the total power drawn from the supplies is constant.

### 12.3.2 Power dissipation in class B amplifiers

For a class B amplifier the quiescent current is set to a (small) value which gives the best transfer characteristic. For the purpose of the analysis this is assumed to be zero. If the load voltage is sinusoidal and has an amplitude $V_{out}$, each output transistor will conduct for half a cycle, and the mean current drawn from either supply will be given by

$$I_{DC} = \frac{V_{out} \int_0^\pi \sin(\omega t)\, d(\omega t)}{R_L \int_0^{2\pi} d(\omega t)}$$

$$= \frac{V_{out}}{\pi R_L} \tag{12.8}$$

and the total power drawn from both supplies will be

$$P_{DC} = \frac{2 V_{out} V_{CC}}{\pi R_L} \tag{12.9}$$

For a class B stage, the power drawn from the supplies is proportional to the output voltage (or the square root of the load power). This is obviously a big advantage over the class A amplifier, where it is independent of the power.

The efficiency is found by dividing the load power by the power from the supplies, and has a maximum value given by

$$\eta = \frac{V_{CC}^2}{2R_L} \cdot \frac{\pi R_L}{2V_{CC}^2} = \frac{\pi}{4} = 0.785 \tag{12.10}$$

when the peak load voltage equals the supply voltage. This is significantly greater than the efficiency of a class A stage, and explains why, except at modest power or where the best linearity is required, the class B output stage is almost universal.

In order to design the heatsink it is necessary to know the maximum power dissipation in the transistors. The maximum instantaneous power will occur when the amplitude of the output signal is $\frac{1}{2}V_{CC}$. In this case equal power will be dissipated in the transistor and the load. Thus the worst case for an a.c. signal would be a square wave of this amplitude, and the power dissipated in each transistor is given by

$$P_T = \frac{1}{2} \cdot \frac{V_{CC}}{2} \cdot \frac{V_{CC}}{2R_L} = \frac{V_{CC}^2}{8R_L} \tag{12.11}$$

However, it is more usual to assume a sinusoidal signal. In this case the power loss in each transistor is found by subtracting the load power given by equation (12.7) from the total power from the supplies given by equation (12.9)and dividing by 2. This gives

$$P_T = \frac{P_{DC} - P_{out}}{2} = \frac{V_{out}}{4R_L} \left( \frac{4V_{CC}}{\pi} - V_{out} \right) \tag{12.12}$$

The maximum value of $P_T$ occurs at $2V_{CC}/\pi$ and is given by

$$P_T = \frac{V_{CC}^2}{\pi^2 R_L} \tag{12.13}$$

This is about 20% less than the square wave 'worst case'.

### 12.3.3  Safe operating area

The calculations of mean power dissipation are suitable for calculating the temperature rise of the heatsink, but it is necessary to ensure that the peak dissipation does not exceed safe limits. At low frequency the junction temperature may vary significantly during the cycle as the power dissipation varies. To ensure safe operation it is necessary to ensure that the voltage/current trajectory lies within the safe operating area (SOA) for the device. With bipolar devices the limiting factor may well be a second breakdown rather than power dissipation.

For a class A stage the transistors are conducting for the whole range of load voltage. Assuming that the quiescent current is set to its minimum value, the current decreases from a maximum value of about $V_{CC/R_L}$ down to zero as the collector-to-emitter voltage increases from zero to $2V_{CC}$. The load line has a slope of $\frac{1}{2}R_L$. The current rating of the transistor must exceed $V_{CC}/R_L$, the voltage rating must exceed $2V_{CC}$, and the load line must remain within the SOA.

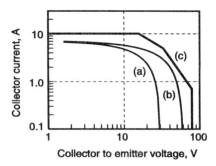

**Figure 12.4** *The locus of the voltage and current for (a) a class B stage, (b) a class A stage, and (c) the d.c. safe operating area*

For a class B stage the current in one of the transistors will decrease from its maximum value of $V_{CC}/R_L$ down to almost zero as the collector-to-emitter voltage increases from zero to $V_{CC}$. The current will then continue to decrease to zero, much more slowly, as the collector voltage rises to $2V_{CC}$. During this second part of the characteristic the current will be very small. The current and voltage ratings for the transistors are as for the class A stage but the load line is different and the power dissipation less.

The load lines for a class A and a class B stage are shown in Figure 12.4, plotted on logarithmic scales, as is usual for SOA diagrams. Also illustrated is the typical shape of the SOA for a bipolar transistor. The figure was produced by simulation of the circuit in Figure 12.1, and assumed a value of 30 V for $V_{CC}$ and a load of 4 $\Omega$. The transistor models used for this simulation are given in Appendix 1, Section A1.8, as models QPWR1 and QPWR2. For the class A amplifier $V_b$ was set to 2.2 V, which gave a quiescent current just sufficient for the amplifier to remain in class A. For the class B amplifier $V_b$ was set to 1.225 V giving a quiescent current of about 10 mA.

## 12.4 Choice of output devices

The output transistors in either a class A or class B stage may be either bipolar transistors or MOSFETs, and both types of device have some merits and some disadvantages. The bipolar transistor has a very large transconductance at high current, which helps to achieve good linearity of the voltage gain for a complementary emitter follower stage with its large local feedback. Modern bipolar power transistors also have adequate bandwidth (an $f_T$ of about 3 MHz, typically) and they are readily available as both $n$–$p$–$n$ and $p$–$n$–$p$ devices with characteristics that are reasonably well matched. At one time selected matched pairs were always used for push–pull stages, While this is desirable, with good circuit design and transistors of reasonable specification it may not be be necessary.

A disadvantage of BJTs is that their charge storage may cause slow turn-off in a class B stage, with consequent high-frequency cross-over distortion. Another problem is their negative temperature coefficient of the base emitter voltage, $V_{BE}$, which, with the large $g_m$, causes poor thermal stability and makes sharing current between parallel devices more difficult.

MOSFETs offer a number of potential advantages for use in power amplifiers. Their transfer characteristics are much more linear than for a bipolar transistor, they are very much faster than bipolar transistors, and, at least at high current, the channel current tends to fall with increasing temperature, avoiding problems of thermal runaway and simplifying current sharing when devices are used in parallel. One problem with MOSFETs for this application is that the characteristics of $p$-channel devices do not match well with $n$-channel devices. This arises from the lower speed of the majority carriers in a $p$-channel device, which requires a much larger device to achieve the same transconductance with a consequent substantial increase in gate source capacitance. The lower value of the transconductance to a large extent offsets their intrinsically more linear transfer characteristics (by permitting less local negative feedback). The larger value of gate source voltage, $V_{GS}$, as compared with $V_{BE}$ for a BJT, means that, using the complementary follower circuit of Figure 12.1, the voltage swing will be reduced if the gate voltage is constrained to swing between the same supply rails as used for the power stage of the amplifier.

Successful amplifier designs are produced using both types of device. BJTs are probably the easier to use, partly because their intrinsically poorer high frequency gain makes it a lot easier to ensure stability and avoid high-frequency oscillation. The larger $g_m$ also makes it easier to achieve good linearity. Bipolar transistors have a small price advantage, given that complementary devices are usually required. For amplifiers intended to work at frequencies above the audio range, MOSFETs would certainly be preferred.

## 12.5   Biasing class A and class B stages

A practical class A or class B output stage requires a method of controlling the bias current. The temperature coefficient of the base emitter voltage for a bipolar transistor is about $-2$ mV $°C^{-1}$, so given that the range of junction temperature might be between 0°C and 125°C, this implies that $V_{BE}$ varies by about 0.25 V for the same collector current. Adequate stabilization of the bias current is essential for an amplifier using BJTs. For a MOSFET the temperature variation is more complex, since the threshold voltage and the transconductance fall with rising temperature, while at high current the channel resistance rises with temperature. At some value of the drain current its temperature coefficient is almost zero. This together with the lower transconductance makes the stabilization of FETs somewhat easier. There is, however, significant variation of threshold voltage between devices of the same type, so adjustment is usually essential.

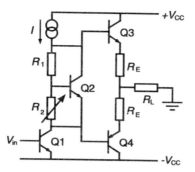

**Figure 12.5** *The Vbe multiplier for providing the bias for a class A or a class B complementary follower*

The usual bias arrangement for complementary emitter follower output stages is shown in Figure 12.5. The bias voltage, $V_b$, is produced using Q2. The voltage drop across Q2 is determined by the ratio of $R_1$ to $R_2$. The circuit is usually referred to as a Vbe multiplier, since the voltage drop, $V_b$, neglecting the base current of the transistor, is given by

$$V_b = \frac{(R_1 + R_2)V_{BE}}{R_2} \tag{12.14}$$

where $V_{BE}$ is the base emitter voltage at the quiescent collector current. The bias voltage, set in this way, will vary in roughly the same way with temperature as the base emitter voltage of the output transistors. To improve the thermal tracking Q2 is often mounted on the heatsink with the power transistors. The matching will be imperfect because the junction temperature may be significantly higher than the heatsink temperature, but it does help to improve the stability of the quiescent current in the output stage.

The other measure taken to improve the stability is the use of the emitter resistors $R_E$. These provide local feedback which reduces the effective $g_m$ of the output transistors, and hence reduces the sensitivity of the quiescent current to changes in $V_{BE}$. The effective value of the transconductance, $g_{m(eff)}$, for a bipolar transistor with an emitter resistor is given by

$$g_{m(eff)} = \frac{g_m}{1 + g_m(1 + 1/\beta)R_E} \tag{12.15}$$

For large values of $g_m$ this reduces to $R_E^{-1}$. The value of $R_E$ is usually kept quite small to avoid excessive power dissipation and to avoid limiting the output voltage swing. For a class B amplifier the value might be in the range 0.3–0.5 $\Omega$. If the desired quiescent current is 50 mA, a value of 0.5 $\Omega$ for $R_E$ would give a value of 1.0 for $g_{m(eff)}$. The stability of the bias current with temperature will not be very good. A 10°C change in junction temperature will produce a change of about 20 mA in the quiescent current. A combination of the use of an emitter resistor with the thermal compensation using the transistor to generate the bias voltage does, in practice, lead to acceptable results, but the design should be able to tolerate changes in the quiescent current without introducing significant distortion.

The same techniques can be used to bias a class A stage. In principle, it should be somewhat easier to ensure that the bias is stable once the circuit is 'warmed up', since the power dissipated in the transistors varies less with the signal power level than it does for a class B stage. The main concern with a class A stage is to ensure that the output transistors are protected against *thermal runaway*. As the junction temperature of a transistor rises, its leakage current increases and its $V_{BE}$ falls. Both these effects increase the collector current and hence the power dissipation, further increasing the temperature. In some circumstances the current and temperature may run away out of control. The criterion for stability is that

$$\frac{\partial P_C}{\partial T} < \frac{1}{R_{\theta ja}} \qquad (12.16)$$

where $P_C$ is the power dissipation in the transistor and $R_{\theta ja}$ is the thermal resistance from junction to ambient (see, for example, Cherry and Hooper, 1968).

In order to illustrate the need to avoid thermal runaway in class A amplifiers, consider an amplifier with supply voltages of $\pm V_{CC}$ and a thermal resistance $R_{\theta ja}$. The power dissipation in the transistor is $I_C V_{CC}$ in the quiescent condition, hence the rate of change of power with temperature is given by

$$\frac{\partial P_C}{\partial T} = \frac{\partial I_C}{\partial T} V_{CC} = -V_{CC} g_{m(eff)} \frac{\partial V_{BE}}{\partial T} \qquad (12.17)$$

Combining equations (12.6) and (12.7) gives the condition for stability:

$$g_{m(eff)} < \left( R_{\theta ja} V_{CC} \frac{\partial V_{BE}}{\partial T} \right)^{-1} \qquad (12.18)$$

For a class A amplifier which delivers 20 W to a load of 8 $\Omega$ a reasonable value of $V_{CC}$ would be about 25 V, the quiescent current would be about 1.2 A, and $R_{\theta ja}$ would need to be about 2°C $W^{-1}$. Taking the temperature coefficient of $V_{BE}$ as $-2$ mV °C$^{-1}$, and substituting these values into equation (12.8) gives the stability condition as $g_{m(eff)} < 10$. The $g_m$ of the transistor at 1.2 A would be about 48 (assuming an ideal transistor), so clearly emitter resistance is needed. A resistance of 0.08 $\Omega$ would be just sufficient to ensure stability; in practice a value of 0.2 to 0.5 $\Omega$ would be preferred. Clearly, thermal runaway is unlikely to be a problem with class B amplifiers where the quiescent current, and hence $g_m$ is much less.

## 12.6 Design considerations for a class B amplifier

The output stage illustrated in Figure 12.5 cannot be operated in isolation. The current in transistor Q1 must be adjusted so that the quiescent output voltage is zero. This is usually achieved by the use of a d.c.-coupled configuration with

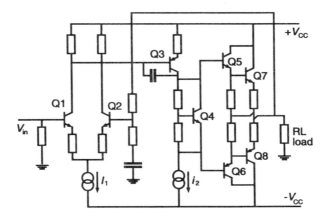

**Figure 12.6** *A simplified circuit for the most popular circuit topology for a class B amplifier*

negative feedback to ensure stability. One of the most popular topologies is illustrated in Figure 12.6. It is, in essence, the same topology as an operational amplifier. There are three stages, two of which provide voltage gain and an output stage to drive the current in the load.

The differential input stage (Q1 and Q2) compensates for the thermal drift of the d.c. levels that would otherwise arise from the variation of the base emitter voltage of the input transistor with temperature. It also provides a convenient method of applying negative feedback to the whole amplifier. A further advantage of the long-tailed pair is that provided it is correctly biased, so that the current in both transistors is equal, the transfer function is symmetrical and provides very good small-signal linearity. This stage generally provides a modest amount of gain (often<20 dB).

The second stage (Q3) provides most of the open-loop voltage gain (>40 dB). The high voltage gain is achieved by the use of a current source active load. The collector current of this stage passes through the Vbe multiplier circuit to generate the bias voltage for the output stage. This stage is also used to provide the dominant pole to ensure stability of the feedback loop when negative feedback is applied to the amplifier. This is usually achieved by the inclusion of a small capacitor across the collector and base junction of Q3. This stage, if not correctly designed, may be a significant source of distortion in the final amplifier, as its output swing must be the full output voltage of the amplifier.

The final stage is the power output stage. The current gain of this stage must generally be in excess of 1000, transforming the signal current from about 1 mA to several amperes. Power transistors normally have rather modest current gain (30–100) at currents in excess of 1 A, so that at least two cascaded circuits are generally required. This may be achieved by the use of two cascaded emitter followers as shown in Figure 12.6. An alternative is to use two complementary feedback pairs as in Figure 12.7. The compound pair of an *n–p–n* transistor driving a *p–n–p*

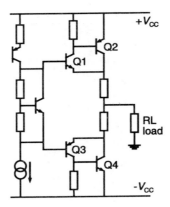

**Figure 12.7** *Complementary pairs in the output stage*

transistor, Q1 and Q2 in Figure 12.7, behaves very much like a single *n–p–n* transistor with very high gain or, of course, a *p–n–p* and an *n–p–n* transistor behave like a single *p–n–p* transistor.

As with a linear regulator, protection must be provided to prevent damage to the output stage in the event of short circuit or an excessive load current being drawn, and is achieved in the same way by limiting the current. The basic current limiting circuit is shown in Figure 12.8. The transistors QP1 and QP2 protect the output stage by diverting the base current if the voltage drop across the emitter resistance of the output transistor exceeds about 0.6 V. This is adequate to protect Q6 and Q8, but to protect Q5 and Q7 the current must also be limited in the second-stage transistor Q3. This is achieved by the use of QP3.

Better protection is achieved by the use of foldback current limiting. This is achieved using a circuit as illustrated in Figure 12.9 for one half of the output stage. If the load voltage is less than or equal to zero the current will limit when the

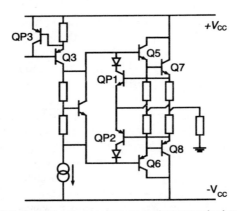

**Figure 12.8** *Protecting the output stage against excessive load current*

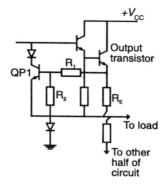

**Figure 12.9** *Providing foldback current limiting for an output transistor*

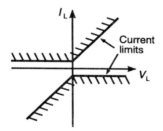

**Figure 12.10** *The current limits for an amplifier with non-linear foldback current limiting*

voltage drop across $R_E$ is sufficient to turn on the transistor QP1. If the load voltage is positive the potential divider formed from R1 and R2 comes into play, and the voltage drop across $R_E$ needed to cause current limiting is increased. Thus using foldback limiting, the maximum current is limited to a low value for low output voltage or output voltage of opposite sign to the current, but the current limit increases with increasing load voltage, as illustrated in Figure 12.10. This limits the power dissipation in the output transistors under fault conditions.

## 12.7 Distortion in audio amplifiers

The output of an audio amplifier should be as closely as possible a replica of the input signal with larger amplitude. In order to achieve this the transfer function must be linear and the gain should be constant across the audio-frequency range, usually taken as about 20 Hz to 20 kHz. The bandwidth is easily characterized, and generally an adequate bandwidth is easily achieved. The more difficult area is linearity. Distortion or deviation from linearity is usually characterised by the *total harmonic distortion* (THD), which is defined by

$$\text{THD} = \sqrt{\frac{P - P_1}{P_1}} \tag{12.19}$$

where $P$ is the total output power and $P_1$ is the power at the fundamental frequency, and it has been assumed that the input signal is a pure tone. This is rather a crude measure, and it does not necessarily indicate how the amplifier will sound, because high-order harmonics are much more noticeable than low harmonics, but it is a useful indication of performance.

Characterizing an amplifier by bandwidth and THD is not, however, complete. If the linearity depends on frequency, or the bandwidth depends on amplitude, as it will in general, the situation is more complex. One type of distortion to which amplifiers which employ feedback are particularly subject is slew rate limitation. If the input signal changes rapidly with time the input stage of the amplifier may overload, and the rate at which the output changes is limited by internal effects. In audio amplifier terms this is usually referred to as *transient intermodulation distortion* or TIM, and its significance has been the subject of much discussion.

The input stage should not introduce significant distortion but the large voltage swing at the output of the second stage and the large current and voltage in the output stage are both potential sources of serious distortion. The output stage is a particular problem for two reasons. In the cross-over region it is impossible to keep the stage transconductance constant, while for large signals the transconductance becomes more constant but the current gain falls because of the falling $\beta$ of the output transistors.

### 12.7.1 Simulation of a class B amplifier

To illustrate the behaviour of a basic power amplifier the circuit shown in Figure 12.11 has been simulated. This is based upon the circuit of Figure 12.6 and is

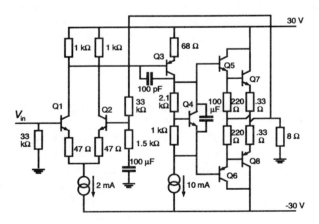

**Figure 12.11** *The circuit diagram for the amplifier simulation*

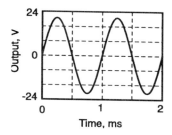

**Figure 12.12**  *The output of the amplifier for a sinusoidal input of amplitude 900 mV*

somewhat simplified to enable it to be simulated using the student version of PSPICE. The netlist is given in Section A1.8 of Appendix 1. The transistor models used do not attempt to represent particular transistor types, but the characteristics have been chosen to represent 'typical' transistors of their type with regard to both their DC characteristics, and their capacitances and transit times.

Transient simulation of an amplifier with a 0.9 V amplitude sinusoidal source gives a clean-looking sinusoidal output (Figure 12.12). Fourier analysis shows that the total harmonic distortion is about 0.037%. The dominant harmonics are the second (0.025%) and the third (0.027%). When simulating such a low-distortion amplifier it is necessary to restrict the time step for the integration to avoid introducing distortion terms due to the numerical integration. Increasing the frequency to 20 kHz increased the distortion to 0.19%. Reducing the load resistance from 8 $\Omega$ to 4 $\Omega$, increased the distortion to 0.12 % at 1 kHz. These low distortion values only arise because of the large amount of feedback (loop gain about 40 dB).

In this simulation the dominant source of the distortion is the falling current gain of the output transistors with increasing current. This is illustrated by plotting the input voltage to the output stage (the average of the voltages at the bases of Q5 and Q6) against the total input current of the output stage (the sum of the base currents of Q5 and Q6). Figure 12.13 shows this input characteristic with an 8 $\Omega$ load. Assuming that all the emitter current of the driver transistor flows into the base of the output transistor, the input impedance of the stage is given approximately by

$$R_{in} \approx \beta_1 \beta_2 R_L \qquad (12.20)$$

where $\beta_1$ is the current gain of whichever driver transistor is conducting, $\beta_2$ is the current gain of the appropriate output transistor and $R_L$ is the load resistance. Also apparent from Figure 12.13 is the phase shift due to the input capacitance. With a 4 $\Omega$ load resistance both the non-linearity and the phase shift become much more apparent (Figure 12.14). The effect of the non-linearity of the input impedance to the output stage is that the voltage gain of the second stage becomes non-linear.

There are several ways of reducing the distortion due to the decrease of the current gain at high load current. One is to use two or more output transistors in parallel, so that each carries a lower current, and obviously to choose transistors

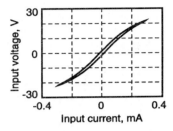

**Figure 12.13** *The input characteristic for the output stage of the simulated amplifier with an 8 Ω load*

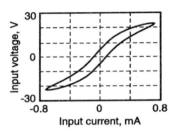

**Figure 12.14** *The input characteristic for the output stage of the simulated amplifier with an 4 Ω load*

that have as constant a current gain as possible. Another option is to reduce the output impedance of the voltage gain stage by reducing the collector load resistance. This will swamp the non-linearity due to the variation of the input impedance of the output stage, at the expense of gain. In the simulation the addition of two 6.8 kΩ resistors between the bases of Q6 and Q7 and ground in the circuit of Figure 12.11 reduced the distortion at a frequency of 1 kHz, with a 4 Ω load, from 0.19% to 0.12%. The addition of these resistors reduces the gain of the amplifier at 1 kHz from 73 dB to 47 dB. Hence, the lower distortion is obtained with a feedback loop gain reduced from 48 dB to 20 dB, implying much better linearity for the open-loop amplifier. Also, the open-loop bandwidth was increased from 1 kHz to about 20 kHz. Interestingly, at 20 kHz the addition of the resistors left the distortion unchanged.

At high frequency (20 kHz) the charge storage effects in the output transistors become a significant source of distortion. This is illustrated in Figure 12.15. Curves (a) and (b) show the base currents of the output transistors when the signal is a 20 kHz sine wave. The negative feedback forces the load voltage (and current) to be sinusoidal. If the system were linear the base currents would each follow half a cycle of a sine wave, but clearly, in the simulation they do not. This is partly due to the falling gain with current. However, the effects of charge storage are clearly visible as asymmetry and a significant reverse base current lasting about 12 μs as the transistors turn off. A transistor cannot turn off until most of this charge has been removed, and this is apparent in curves (c) and (d) which show the emitter currents. In the cross-over region there is a significant overlap, where both transistors are conducting.

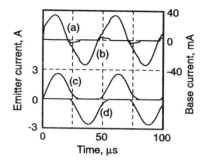

**Figure 12.15** *The simulated current flowing in the output transistors at a frequency of 20 kHz. (a) Base current of Q7; (b) base current of Q8; (c) emitter current of Q7; (d) emitter current of Q8*

Only a few of the sources of distortion in a power amplifier have been explored here. In practice, to produce a low-distortion amplifier calls for attention to the details of the circuit design, the layout and the choice of the passive and active components. A series of articles by Self (1993, 1994) examines a wide range of mechanisms producing distortion, and there are some interesting comments on this series by Cherry (1995).

## 12.8  Class D amplifiers

As discussed in earlier chapters, switching techniques using pulse-width modulation have become common in the control of power for both a.c. and d.c.. The same techniques described when discussing inverters with sinusoidal PWM can be applied to amplifiers, with the difference that the output voltage to be synthesized contains a large number of frequency components spread over a wide range of frequency and amplitude. Amplifiers operating in this way are usually described as class D. The advantage to be gained is that of high conversion efficiency from d.c. to the audio signal.

Unipolar modulation, as discussed in Chapter 10, could be used, i.e. the pulses of variable width are either positive or negative, depending on the sign of the output signal. Such a scheme would be unsuitable for an audio amplifier as cross-over distortion when switching from a positive to negative signal would be very difficult to avoid. Hence for class D amplifiers a bipolar scheme is universally employed, and the output is switched from positive to negative at high frequency as shown in Figure 12.16.

Figure 12.17 shows a block diagram of a class D amplifier. The oscillator generates an accurate triangular wave which is compared with the input signal to generate the pulse-width modulation. The comparator drives the output stage, which switches the load voltage between fixed voltages. The output is filtered using a multistage filter with an inductive input impedance at the switching frequency. To

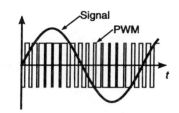

**Figure 12.16** *Pulse-width modulation*

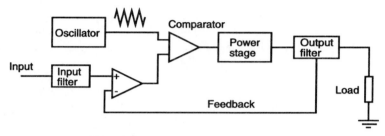

**Figure 12.17** *A block diagram of a class D amplifier*

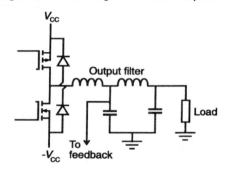

**Figure 12.18** *A class D output stage*

reduce distortion negative feedback is applied by subtracting a fraction of the output voltage from the analog input signal.

The output stage of a class D amplifier generally uses a half-bridge as illustrated in Figure 12.18, and its topology is the same as for an inverter. The filter is critical in that it must provide a suitable inductance to limit the current at the switching frequency. It must have a bandwidth of about 20 kHz, and it must provide a large attenuation at the switching frequency to avoid RF emissions. To avoid excessive phase shift, the feedback signal may be taken part of the way through the filter.

The switching frequency needs to be high, around 500 kHz, for a 20 kHz bandwidth. Thus MOSFETs are the preferred switches. The switch timing must be precise. The turn-on and turn-off delays of the switches must be independent of the load current and should be as short as possible if distortion is to be avoided. To avoid excessive power loss during switching, the switches need to have a short rise and fall time. However, overlap of the switches as they turn on and off must be avoided. Introducing a significant dead-time will introduce distortion, so matching of turn-on and turn-off times is essential. The inherent asymmetry between $n$-channel and $p$-channel MOSFETs makes this matching difficult if complementary devices are used. If only $n$-channel devices are used then the problem is one of isolating the gate drive for the top device while maintaining the speed.

Class D amplifiers, despite their promise of high efficiency, have not become popular. They do present the designer with many challenges, and it is probable that the complexity is not justified unless a very large output power is required. The high switching frequency required to obtain low distortion across the whole audio bandwidth is the major source of difficulty. Where a narrower bandwidth can be tolerated and a switching frequency of around 100 kHz can be accepted the problems become less severe.

## Summary

This chapter has examined some of the issues concerned with the design of audio-frequency power amplifiers, paying particular attention to output stages. Class A and class B amplifiers have been analyzed to determine their power dissipation and efficiency. The design of class B amplifiers based on the complementary emitter follower output stage has been examined and simulation used to illustrate some of the sources of distortion. Finally, class D amplifiers, using switching techniques and PWM, have been considered briefly.

## Tutorial questions

**12.1** Explain the role of the emitter resistors in a class B output stage using the complementary emitter follower topology.

**12.2** Why does a class A push–pull output stage usually have a lower distortion than a class B stage?

**12.3** Why for a class A stage does the power dissipated in the output transistors fall as the load power rises?

**12.4** Show that for a class B stage the instantaneous power dissipation is greatest when the load voltage is half the supply voltage, $V_{CC}$, assuming split supplies.

**12.5** Why is second breakdown less likely to be a problem with the output stage of a class B amplifier as opposed to a class A amplifier?

**12.6** What are the relative merits of BJTs and MOSFETs when used as output stages in class A or class B amplifiers?

**12.7** What are the principal sources of distortion in a class B amplifier?

**12.8** Explain how a class D amplifier is able to achieve an efficiency of 90%.

**12.9** What are the principal problems with a class D amplifier?

# References

Cherry, E. M., 'Ironing out distortion', *Electronics and Wireless World*, 14–20, January, 1995.

Cherry, E. M. and Hooper, D. E., *Amplifying Devices and Low-pass Amplifier Design*, John Wiley, New York, 1968.

Self, D., 'Distortion in power amplifiers', *Electronics and Wireless World*, August 1993 to March 1994.

# *SPICE* listings

This appendix contains the listings for the SPICE simulations used in the main text. All these listings have been used with PSPICE. If other versions of SPICE are used then some minor changes may be required.

## A1.1  Full-wave rectifier with reservoir capacitor

```
* simulation of power supply
* Source
V1  1  2  0  SIN(0,14.14,50)
* source resistance
RS  1  3  0.2
* Diode bridge
D1  3  4  diode
D2  2  4  diode
D3  0  3  diode
D4  0  2  diode
* Filter
CF  4  0  4700uF
* Load
RL  4  0  16

.MODEL diode D(IS=10fA N=0.01 TT=100ns CJO=50pF)
.TRAN 100u  200m  180m  100u  UIC
.PRINT TRAN V(4)  V(6)  V(7)  I(RS)  I(RSa)
.PROBE
.END
```

## A1.2  Full-wave rectifier with inductance input filter

```
* simulation of power supply with inductive input filter
* Source
```

```
V1  1  2  0  SIN(0,35.35,50)
* Diode bridge
D1  1  3  diode
D2  2  3  diode
D3  0  1  diode
D4  0  2  diode
* Filter
RS  3  4  0.001
LF  4  5  50mH
CF  5  0  4700uF
* Load
RL  5  0  22

.MODEL diode D(IS=10fA N=0.01)
.TRAN 1m, 500m, 0, 1m UIC
.PRINT TRAN V(5)  I(LF)
.PROBE
.END
```

## A1.3  Series regulator

```
* Linear regulator with foldback current limiting
R1  1  2  0.12
R2  1  3  10k
R3  2  4  270
R4  2  5  100
R5  4  9  10k
R6  1  6  1k
R7  8  0  390
R8  9  7  10k
R9  7  0  4.7k
C1  9  0  10uF
RL  9  10  1              * load resistance
VS  1  0  20V             * source voltage
VL  10  0  14.5V pulse(15, -2, 1E-3, 1, 1, 1, 4)     * to
sweep load current
Q1  9  5  2  Qpass
Q2  9  6  5  Qdrv
Q3  6  4  1  Qlim
Q4  6  3  8  Qamp
Q5  1  7  8  Qamp
D1  0  3  Z4-7
* device models
```

```
* Zener diode
.MODEL Z4-7 D(Is=1u Rs=10 Bv=4.6 Ibv=1u)
* differential amp transistor
.MODEL Qamp NPN (Is=50f Xti=3 Eg=1.11 Bf=650 Br=10 Vaf=80
+               Vjc=.84 Cjc=3.8p Mjc=.314 Vje=.84 Cje=7.0p
Mje=.4
+               Tr=100n Tf=500p Rb=200)
* current limiting transistor
.MODEL Qlim PNP  (Is=50f Bf=200 Vaf=80 Br=10 Var=10 Rb=200
+               Cje=12p Vje=.75 Mje=.3 Cjc=7p Vjc=.7
Mjc=.425
+               Tf=400p Tr=100p)
* driver transistor
.MODEL Qdrv PNP  (Is=50f Bf=100 Br=5 Vaf=70 Var=10
+               Cje=200p Mje=.35 Vje=.7 Cjc=100p Mjc=.35
+               Vjc=.6 Tf=2n Tr=100n Rb=100)
* pass transistor
.MODEL Qpass PNP (Is=2p Bf=50 Vaf=70 Br=5 Var=10 Rb=50
+               Cjc=300p Mjc=.35 Vjc=.55 Cje=350p Mje=.4
Vje=.7
+               Tf=50n Tr=.5u)

.tran  .01  1  0  0
.probe
.END
```

## A1.4  Pulse-width modulator

```
* pwm control of power in an inductive load
* Source voltage
Vs  1  0  50V
* Power switch
M1  1  2  4  4  mosfet
R1  2  3  25
* Load
L1  4  5  2mH
R2  5  0  25
* Freewheeling diode
D1  0  4  Diode
* Control voltage for switch
Vg  3  4  0  PULSE(0 10 0 5n 5n 5u 10u)
.model mosfet    NMOS (Level=3 Tox=100n Uo=600 Phi=.6 Rs=50m
+ Kp=20u W=.15 L=2u Vto=2.7
```

```
+                       Rd=.2 Cgso=3n Cgdo=1n Rg=50)
.model Diode D(IS=1p CJO=50pF)
.tran 100n 100u
.probe
.print tran V(4) V(5)
.end
```

## A1.5  Model for the feedback loop of a buck regulator

```
* Model of buck regulator feedback amplifier, open loop

R1  1  2  4.7k
R2  2  0  4.7k
R3  3  0  5Meg
R4  4  0  33k
R5  7  0  100m
RL  6  0  5

C1  3  0  106pF
C2  3  4  15nF
C3  6  7  470uF

L1  5  6  100uH
* controlled sources
G1  3  0  2  1  0.002S
E1  5  0  3  0  5.7
* AC excitation
V1  1  0  0  AC 1

.ac dec 50 1 100k
.probe
.end
```

## A1.6  Half-controlled rectifier with an R-L-e.m.f. load

```
* Simulation of a half-controlled converter
.PARAM delay = 3.355m period=20m switch_time=8m

V1  1  2  0  SIN(0 155 50)
L1  1  3  100uH
Rs  1  3  100
X1  3  4  SCR  PARAMS: td={delay} tsw={switch_time}
X2  0  3  SCR  PARAMS: td={delay+period/2}
```

```
tsw={switch_time}
D1   0   2   diode
D2   2   4   diode
LL   4   5   100mH IC=9.5A
RL   5   6   2
VB   6   0   54V


.model diode D(IS=1pA Cjo=100pF TT=1us)

.subckt SCR  1  3  PARAMS: td=1m tsw=10m
S1   1   2   4   3   switch
Rs   1   5   200
Cs   5   3   0.1uF
D1   2   3   dtype
R1   4   3   1k
Vt   4   3   0   PULSE(0 1 {td} 1u 1u {tsw} {period})
.model dtype D(IS=1p cjo=100pF TT=10us)
.model switch VSWITCH (RON=0.01 ROFF=10MEG VON=0.7 VOFF=0.2)
.ends

.tran 200u 40ms uic
.probe
.print tran V(4) I(VB) I(V1)
.four 100 V(4) I(VB) I(VI)
.END
```

## A1.7  Current source inverter

```
* Sinusoidal PWM, current-source inverter
I1   3   1   10A
Cs   2   3   0.5uF
Rs   1   2   10
S1   1   4   9   7   smod
S2   5   3   7   0   smod
S3   1   6   7   10  smod
S4   8   3   0   7   smod
D1   4   0   dmod
D2   0   5   dmod
D3   6   7   dmod
D4   7   8   dmod


VL   7   0   0.2  sin(0.2 340 50)
RL   7   0   1Meg
```

```
* Sawtooth reference voltages
VR1  9   0   0   pulse(0 360 0 499.5u 499.5u 1u 1m)
VR2  10  0   0   pulse(0 -360 0 499.5u 499.5u 1u 1m)
RR1  9   0   1Meg
RR2  10  0   1Meg
.model smod VSWITCH (Ron=0.1 Roff=1MEG Von=0.1 Voff=-0.1)
.model dmod D(IS=1p)
.tran 1m 20m
.probe
.options ITL5=20000 abstol=0.01 reltol=.01 vntol=0.1
.end
```

## A1.8   Class B amplifier

```
* Class B amplifier
R1    1    0    33k
R2    40   2    1k
R3    3    4    47
R4    40   5    1k
R5    6    4    47
R6    4    30   10k
R7    7    8    1.5k
R9    40   9    68
R10   10   11   2.1k
R11   11   12   1k
R13   15   16   220
R14   16   17   220
R15   18   16   0.33
R16   16   19   0.33
R1    16   0    8
Rf    16   7    33k
R20   20   0    1k

C1    8    0    10mF
C2    20   1    10uF
C3    2    10   100pF
C4    10   12   100uF
* Power supplies
VC1   40   0    30
VC2   0    30   30
* Sinusoidal source voltage
Vs    20   0    0    AC  1  SIN(0 900m 1k)
* Dummy voltage sources for measuring current
```

```
vm1   21   10   0
vm2   22   12   0
* Bias current source
I1   12   30   10mA
* Active devices
Q1    2    1    3   Qnpn
Q2    5    7    6   Qnpn
Q3   10    2    9   Qpnp
Q5   10   11   12   Qnpn
Q6   40   21   15   QDRV1
Q7   30   22   17   QDRV2
Q8   40   15   18   QPWR1
Q9   30   17   19   QPWR2
* Device models for the transistors
.MODEL QNPN NPN  (Is=30f Bf=10 Ne=1.3 Isw=20f Ikf=.12
+                Vaf=80 Var=0.5 Vjc=.84 Cjc=3.8p Mjc=.3
+                Vje=.84 Cje=7.0p Mje=.4 Tr=100n Tf=500p
+                Rb=200 Rd=1)
.MODEL QPNP PNP  (Is=30f Bf=330 Br=10 Ne=1.3 Isw=20f Ikf=.12
+                Vaf=80 Var=9.5 Vjc=.84 Cjc=3.8p Mjc=.3
+                Vje=.84 Cje=7.0p Mje=.4 Tr=100n Tf=500p
+                Rb=200 Re=1)
.MODEL QDRV1 NPN (Is=50f Bf=150 Br=5 Ikf=.6 Vaf=75 Var=10
+                Ise=100f Ne=1.4 Cje=225p Mje=.35 Vje=.7
+                Cjc=100p Mjc=.35 Vjc=.7 Tf=2n Tr=100n
+                Rb=100 Re=.1)
.MODEL QDRV2 PNP (Is=50f Bf=150 Br=5 Ikf=.6 Vaf=75 Var=10
+                Ise=100f Ne=1.4 Cje=225p Mje=.35 Vje=.7
+                Cjc=100p Mjc=.35 Vjc=.7 Tf=2n Tr=100n
+                Rb=100 Re=.1)
.MODEL QPWR1 NPN (Is=20p Bf=110 Vaf=75 Ikf=8 Ise=800f Ne=1.2
+                BR=5 Var=10 Rb=1 Cjc=125p Mjc=.35 Vjc=.7
+                Cje=290p Mje=.35 Vje=.7 Tf=50n Tr=0.5u)
.MODEL QPWR2 PNP (Is=20p Bf=110 Vaf=75 Ikf=8 Ise=800f Ne=1.2
+                Br=5 Var=10 Rb=1 Cjc=125p Mjc=.35 Vjc=.7
+                Cje=290p Mje=.35 Vje=.7 Tf=50n Tr=0.5u)

* Transient analysis
.tran  25u 2.000m 0 5u
.print  tran v(16) v(10) v(12) I(vm1) I(vm2)
.probe
* Fourier analysis of output voltage
.four  1.000k V(16)
* Small signal analysis to obtain frequency response
```

```
.ac dec 10  10  1.000meg
.print  ac V(16)
.op
.end
```

# Answers to self-assessment questions

## Chapter 2

2.1 Thermal resistance = 0.33°C/W
2.2 Junction temperature = 71.9°C
2.3 Maximum thermal resistance = 1.75°C/W
2.4 (a) Maximum power = 16.7 W
    (b) Maximum power = 22.6 W
2.5 Time to reach 125°C = 11.4 s
2.6 Maximum thermal resistance = 0.67°C/W

## Chapter 3

3.1 Reluctance = $1.59 \times 10^5\,H^{-1}$
3.2 Reluctance = $4.14 \times 10^6\,H{-}1$
3.3 Inductance = $151\,\mu H$
    Maximum current = 11.7 A
3.4 Inductance factor = $4.67\,\mu H/turn^2$
3.5 Maximum stored energy = 1.0 mH
3.6 Primary turns = 71
    Secondary turns = 36

## Chapter 4

4.1 $V_{rrm} > 142\,V$
    $I_{F(av)} > 5\,A$
4.2 Conduction angle = 0.254 rad
    Peak-to-peak ripple voltage = 9.19 V
4.3 Peak input current = 2.38 A
    R.M.S. value of source current = 0.416 A
4.4 Average current = 0.283 A
    R.M.S. diode current = 1.71 A
4.5 Critical load resistance = 113.1 Ω
    Critical load current = 1.75 A

Peak-to-peak ripple voltage = 0.93 V

4.6

$$\text{Ripple current} = \frac{4\sqrt{2}V}{\pi(n^2 - 1)n\omega L}$$

4.7   VA product = 444 VA
Power factor = 0.32

4.8   Peak-to-peak ripple $\simeq$ 8.33 V

4.9   Source voltage = 71 V r.m.s.
Capacitor = 200 $\mu$F
Capacitor current = 0.667 A r.m.s.

# Chapter 5

5.1   Maximum source resistance = 40 $\Omega$
(a) Maximum power dissipated in resistance = 5.63 W
(b) Maximum power dissipated in shunt regulator 5.63 W

5.2   Stabilization factor = 0.053
Output resistance = 9.8 $\Omega$

5.3   Line regulation = 2.2%
Load regulation = 1.0%

5.4   (a) Maximum power dissipated in pass transistor = 20 W
(b) Maximum power dissipated in pass transistor = 28 W

# Chapter 6

6.1   Average current = 2.5 A

6.2   Peak-to-peak ripple current = 37.5 mA

6.3   Mean load current unchanged
Peak-to-peak ripple current = 1.85 A

6.4   Switching loss = 4 W

6.5   Switch total power loss = 4.4 W
Diode power loss = 0.48 W

6.6   Maximum rate of rise = 10 A/$\mu$s
Minimum off-time = 2.5 $\mu$s

6.7   Maximum switch voltage = 72.5 V
Power dissipated in resistor = 1.27 W

# Chapter 7

7.1   Minimum mean load current = 0.21 A

7.2   Minimum inductance $\simeq$ 334 $\mu$H

7.3   Minimum load current for continuous conduction = 0.22 A
7.4   Duty ratio = 0.258
7.5   Maximum load current = 37.5 mA
7.6   Energy stored = 25 $\mu$J
7.7   Maximum on-time = 6.25 $\mu$s
      Maximum inductance = 17.6 $\mu$H
7.8   Maximum duty ratio = 0.5
      Maximum load voltage = 20 V
      Maximum switch-voltage = 200 V

## Chapter 9

9.1   Firing angle = 1.23 rad (70.6°)
9.2   Mean load current 9 A. Yes, it is reasonable!
9.3   Amplitude of ripple current = 1.01 A
9.4   PF = 0.840
      HF = 0.957
9.5   (a) Overlap angle = 38.6 mrad
      (b) Overlap lasts 123 $\mu$s

## Chapter 10

10.1  Peak current = 3.81 A
10.2  R.M.S. value of fundamental = 56.6 V
      R.M.S. fundamental current = 1.52 A
      Ripple current may be neglected
10.3  (a) R.M.S. load voltage = 155 V
      (b) R.M.S. of fundamental = 145.5 V
10.4  R.M.S. of load voltage = $V_S\sqrt{D}$
10.5  Peak load voltage = 27.7 V
10.6  Voltage zero occurs at 0.44 ms after current changes sign
10.7  Time between pulses = 3.28 ms
10.8  R.M.S. magnitude of third harmonic = 12.7 V
      R.M.S. magnitude of fifth harmonic = 27 V

## Chapter 11

11.1  Current = 1.33 A
      Voltage = 10.28 V
11.2  (a) Duty ratio = 0.685
      (b) Duty ratio = 0.843

11.3  Ripple current = 50 mA peak-to-peak

11.4  (a) Resistance = 0.5 Ω
           Time constant = 4.7 ms
           Power = 6.1 W
      (b) Resistance = 5.66 Ω
           Time constant = 1.17 ms
           Power = 69 W

11.5  Time to reverse current = 2.36 ms

# Index